# 5th Young Researcher Meeting – Trieste 2014



**Journal of Physics: Conference Series Volume 566**

## Trieste, Italy
## 14 – 15 July 2014

**ISBN: 978-1-63439-752-0**
**ISSN: 1742-6588**

**Printed from e-media with permission by:**

Curran Associates, Inc.
57 Morehouse Lane
Red Hook, NY 12571



Printed by Curran Associates, Inc. (2015)

For permission requests, please contact the Institute of Physics
at the address below.

Institute of Physics
Dirac House, Temple Back
Bristol BS1 6BE UK

Phone: 44 1 17 929 7481
Fax: 44 1 17 920 0979

techtracking@iop.org

**Additional copies of this publication are available from:**

Curran Associates, Inc.
57 Morehouse Lane
Red Hook, NY 12571 USA
Phone: 845-758-0400
Fax: 845-758-2634
Email: curran@proceedings.com
Web: www.proceedings.com

# TABLE OF CONTENTS

Journal of Physics: Conference Series **566** (2014) 011001 doi:10.1088/1742-6596/566/1/011001

# Young Researcher Meeting, Trieste 2014

F Agostini, C Antolini, R Aversa, G Cattani, M Di Stefano, M Longobardi, M Martinelli, A Miceli, M Migliaccio, F Paci, D Pietrobon, E Pusceddu, F Stellato

## PREFACE

The Young Researcher Meeting (www.yrmr.it) has confirmed once again this year the enthusiasm and determination of Ph.D. students, postdoctoral fellows and young researchers to play a major role in the scientific progress. Since 2009, we regularly gather together to discuss the most recent developments and achievements in Physics, firmly convinced that sharing our expertise and experience is the foundation of research activity.

The format we chose is an informal meeting primarily aimed at graduate students and postdoctoral researchers, who are encouraged to present their work in brief presentations that provide genuine engagement of the audience and cross-pollination of ideas. One of the main purposes of the conference is to create an international network of young researchers, both experimentalists and theorists, and fruitful collaborations across the different branches of Physics.

After four editions that strengthened it, the Young Researcher Meeting 2014 was held at the International School for Advanced Studies – SISSA, Trieste, for the second time. The fifth appointment was a two-day conference on July $14^{th}$-$15^{th}$ 2014. It has been sponsored by a number of research groups of SISSA, the University of Udine and the Solar Physics group of the University of Rome "Tor Vergata", thus gathering even broader support than previous editions. The success of this year event is testified by an increased number of participants and institutions all around Europe. This resulted in an extremely rich and interactive poster session that covered several areas of pure and applied Physics.

With the intent of broadening the contents and stimuli adopting multidisciplinary tools, the YRM 2014 hosted the workshop "Communicating Science" held in collaboration with SISSA Medialab. This choice reflects the growing importance of the outreach activity performed by scientists, especially at the earliest stages of their career, as a way of increasing their expertise and developing soft skills. Engaging the public and finding unconventional ways to communicate results turn out to be real assets in improving the quality of presentation of current research to peers, as well as to the general public.

In this volume, we collect part of the contributions that have been presented at the conference. They cover topics in astrophysics and cosmology, particle and theoretical physics, soft and condensed matter, medical physics and applied physics. Given the recent experimental achievements in particle physics and cosmology, several contributions were focused on the latest results obtained in these fields, presenting the impact of experiments such as LHC and Planck to the community of young researchers and forecasting the future goals in these areas of research. Particular interest was aroused by the session fully dedicated to applied Physics and conservation of cultural assets. Besides the intrinsic scientific value of the discussed topics, the increasing relative weight of the applied Physics session is a demonstration of the benefits that fundamental science brings to the community.

## YRM Organising and Editorial Committee

**Fabio Agostini**
(fabioagostini31@gmail.com)
Telespazio A Finmeccanica Thales Company

**Claudia Antolini**
(claudia.antolini@sissa.it)
SISSA – Scuola Internazionale Superiore di Studi Avanzati and Fudan University

**Rossella Aversa**
(raversa@sissa.it)
SISSA – Scuola Internazionale Superiore di Studi Avanzati

**Giordano Cattani**
(giordano.cattani@gmail.com)

**Marco Di Stefano**
(distefan@sissa.it)
SISSA – Scuola Internazionale Superiore di Studi Avanzati

**Maria Longobardi**
(marialongobardi@gmail.com)
Department of Condensed Matter Physics, University of Geneva

**Matteo Martinelli**
(martinelli@thphys.uni-heidelberg.de)
SISSA – Scuola Internazionale Superiore di Studi Avanzati and Institut fur Theoretische Physik

**Alice Miceli**
(alice.miceli@uniroma2.it)
Physics Department, University of Rome Tor Vergata

**Marina Migliaccio**
(mm858@ast.cam.ac.uk)
Institute of Astronomy and Kavli Institute for Cosmology, University of Cambridge

**Francesco Paci**
(fpaci@sissa.it)
SISSA – Scuola Internazionale Superiore di Studi Avanzati

**Davide Pietrobon**
(davide.pietrobon@berkeley.edu)
University of California at Berkeley

**Emanuela Pusceddu**
(emanuela.pusceddu@gmail.com)
Institute of Biometeorology CNR

**Francesco Stellato**
(francesco.stellato@roma2.infn.it)
INFN Roma Tor Vergata

Journal of Physics: Conference Series **566** (2014) 011001 doi:10.1088/1742-6596/566/1/011001

## ACKNOWLEDGMENTS

The organisers of the 5th Young Researcher Meeting would like to thank all the scientists who participated to the meeting. We furthermore thank all our sponsors that are listed below for supporting the event. We are grateful to the International School for Advanced Studies (SISSA) for hosting the conference for the second time, and to its director, Prof. Guido Martinelli, for his support and advice. We owe gratitude to SISSA Medialab, for organising the public event on science communication and providing technical support throughout the entire meeting. The publication of the proceedings of the conference is partially supported by the Solar Physics group in Tor Vergata; we also acknowledge support from the University of Udine.

The event was broadcast live by OggiScienza (http://oggiscienza.wordpress.com). The complete videos of the meeting can be found at the YRM Youtube channel

https://www.youtube.com/channel/UCw3roeK9oC4NPc-sRQ2t0rg

International School for Advanced Studies (SISSA), Trieste

PRIN 2010-2011 (MIUR 2010YJ2NYW_001) - "Symmetries, Masses and Mysteries: Electroweak symmetry breaking, flavor mixing and CP violation, and Dark Matter in the LHC era" - SISSA, Trieste

Molecular and Statistical Biophysics Group - SISSA, Trieste

PRIN 2012 (2012CPPYP7_006) - "Theoretical Astroparticle Physics" - SISSA, Trieste

PRIN 2010-2011 (MIUR 2010NHBSBE_008) - "L'Universo oscuro e l'evoluzione cosmica dei barioni: dalle survey attuali a Euclid" - SISSA, Trieste

Department of Chemistry, Physics and Environment of the University of Udine

Solar Physics Group - Department of Physics of the University of Rome "Tor Vergata"

Journal of Physics: Conference Series **566** (2014) 011002 doi:10.1088/1742-6596/566/1/011002

# Peer review statement

All papers published in this volume of *Journal of Physics: Conference Series* have been peer reviewed through processes administered by the proceedings Editors. Reviews were conducted by expert referees to the professional and scientific standards expected of a proceedings journal published by IOP Publishing.

5th Young Researcher Meeting, Trieste 2014 IOP Publishing
Journal of Physics: Conference Series **566** (2014) 012001 doi:10.1088/1742-6596/566/1/012001

# Searches for direct stop production within the ATLAS experiment

**Paolo Dondero** *on behalf of the ATLAS Collaboration*
Universit`a degli Studi di Pavia, INFN Pavia E-mail: paolo.dondero@cern.ch

**Abstract.** The ATLAS experiment at the LHC, in conjunction with the discovery of the Higgs boson is looking for signs of physics which go beyond the Standard Model of Electroweak interactions. Among possible theories for physics beyond Standard Model, Supersymmetry seems to be the most promising one. This theory indeed addresses the Standard Model naturalness problem and offers a perfect candidate for the dark matter. Within this scenario the search for a supersymmetric partner of the top quark, called stop, plays a key role. The ATLAS Experiment has developed a dedicated strategy for the discovery of this particle, focusing on achieving a complete coverage of the available parameter space for this particle, based on the combined search for all of its possible decay modes. The results obtained using the complete ATLAS 2012 statistics will be presented, targeting different decay modes and explaining the procedure to obtain the exclusion limits on the existence of a supersymmetric partner of the top quark at the electroweak scale. [1]

## 1. Introduction

Supersymmetry (SUSY) is an extension of the Standard Model (SM) which introduces partners (called superpartners) for each of the known Standard Model bosons and fermions in order to solve open issues in the SM such as the hierarchy problem [1].

In order to avoid violation of baryon and lepton numbers in the theory a new multiplicative absolutely conserved number is introduced, defined as $R = (-1)^{3(B-L)+2S}$ and called R-parity. In the framework of a generic R-parity conserving minimal supersymmetric extension of the SM, the MSSM (Minimal Supersymmetric Standard Model), SUSY particles are produced in pairs and the lightest particle is stable (LSP). In multiple models this particle is the lightest neutralino $\tilde{\chi}_1^0$, that is also a natural candidate for dark matter.

Arguments related to the quantum stability of the mass of the Higgs boson suggest that the SUSY partners of the top (stop) and the bottom (sbottom) quarks have masses below the TeV scale. In this situation the direct production of a stop pair can be observed at LHC in the ATLAS detector, a general-purpose experiment described in details in Ref. [7]. An extensive search program has been developed to this purpose.

In these proceedings the latest results on direct stop production searches within the ATLAS experiment are interpreted and summarized. Among the many analyses published to date on this topic, in the following sections two stop searches, both requiring two leptons in the final state, will be presented in order to illustrate the procedure used in ATLAS to explore a given stop squark decay topology and phase space region. At the end a brief overview of the current ATLAS results will be given.

## 2. Event selection

The two searches share a common preselection and some discriminatory event-level variables. The following variables are defined:

- $m_{ll}$: the invariant mass of the two final state leptons
- $m_{T2}^{ll}$ and $m_{T2}^{b-jet}$: a generalization of the invariant mass [5, 6] defined as

$$m_{T2}^2(\vec{p}_T^{\alpha}, \vec{p}_T^{\beta}, \vec{p}_T^{miss}) = \min_{\vec{q}_T^1 + \vec{q}_T^2 = \vec{p}_T^{miss}} \left[ \max(m_T^2(\vec{p}_T^{\alpha}, \vec{q}_T^1), m_T^2(\vec{p}_T^{\beta}, \vec{q}_T^2)) \right]$$

  where $m_T$ indicates the transverse mass, $\vec{p}_T^{\alpha}$ and $\vec{p}_T^{\beta}$ are the transverse momenta of two particles, $\vec{q}_T^1$ and $\vec{q}_T^2$ are vectors which satisfy $\vec{q}_T^1 + \vec{q}_T^2 = \vec{p}_T^{miss}$. The minimisation is performed over all the possible decompositions of $q_T$. If $\alpha$ and $\beta$ are taken as the two leptons in the final state of the event (indicated with $m_{T2}^{ll}$, simply $m_{T2}$ in the following) the distribution is bound sharply from above by the mass of the W. Selecting events with $m_{T2} > m_W$ is thus possible to enhance the signal over background ratio.
- $\Delta\phi_j$: the azimuthal angle difference between $p_T^{miss}$ and the direction of the closest jet.
- $\Delta\phi_b$: the azimuthal angle between $p_T^{miss}$ and $p_{Tb}^{ll} = p_T^{miss} + p_T^{l1} + p_T^{l2}$.
- $E_T^{miss}$: missing transverse energy in the event.

The selected events are required to have two leptons (electrons or muons) opposite sign. The leading selected lepton must have transverse momentum greater than 25 GeV and the dilepton invariant mass $m_{ll}$ greater than 20 GeV.

## 3. Leptonic $m_{T2}$ analysis

The stop squark can decay into several final states depending in particular on the hierarchy of the mass eigenstates set up from the mixing of the Higgs and gauge bosons superpartners. The eigenstates addressed by the present analysis are the lightest chargino ($\tilde{\chi}_1^{\pm}$) and the neutralino ($\tilde{\chi}_1^0$), supposed to be the LSP in this scenario.

To study the stop squark decay in the dileptonic channel three different approaches have been developed, depending on the mass hierarchy of the involved sparticles: the leptonic $m_{T2}$, the hadronic $m_{T2}$ and the multivariate (MVA) analyses (figure 1, respectively a,b and c). These searches address events with two isolated leptons (electrons or muons) with opposite charge, two b-quarks and significant missing transverse momentum ($E_T^{miss}$) in the final state [2].

In particular the leptonic $m_{T2}$ analysis is designed to focus on the stop decay channel $\tilde{t} \to b + \tilde{\chi}_1^{\pm}$ in the case $m(\tilde{\chi}_1^{\pm} - \tilde{\chi}_1^0) > m_W$, so the $\tilde{\chi}_1^{\pm}$ can decay into the LSP and a W boson.

After base preselections, events with same flavour leptons are required to have the invariant mass of the pair outside the 71 - 111 GeV range in order to reduce the number of background events containing two leptons due to the decay of a Z boson.

Two additional cuts are applied to reduce events with high $m_{T2}$ arising from events with large $E_T^{miss}$ due to an error in the jets measure: $\Delta\phi_b < 1.5$ and $\Delta\phi_j > 1$.

The expected dominant Standard Model background contributions in the signal regions (SRs) are $\bar{t}t$ and dibosons decays. These contributions are normalized to data using control regions (CRs) where each background source is dominant with respect to signal. Dedicated control regions are defined for $\bar{t}t$, WW and Z (for WZ and ZZ).

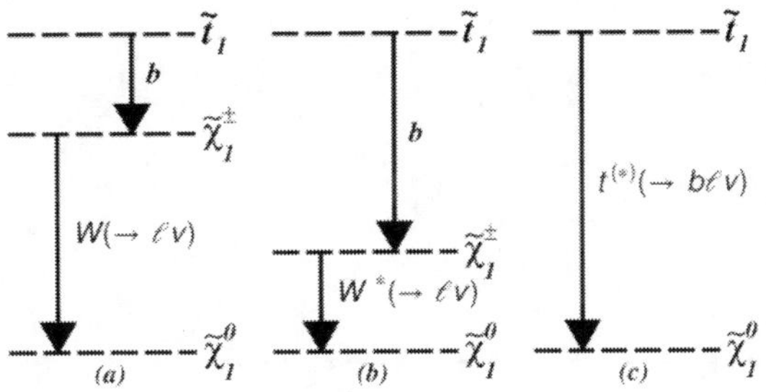


**Figure 1.** Stop decays channels addressed by the leptonic $m_{T2}$ (a), hadronic $m_{T2}$ (b) and MVA (c) analises. The leptonic $m_{T2}$ analysis focuses on the case $\tilde{t} \to b + \tilde{\chi}_1^\pm$ with $m(\tilde{\chi}_1^\pm - \tilde{\chi}_1^0) > m_W$. The figure is taken from Ref. [2].

Four signal regions are defined with different selections on $m_{T2}$ and on the transverse momentum of the two leading jets, sensitive to different configuration of $\Delta m(\tilde{t}, \tilde{\chi}_1^\pm)$ and $\Delta m(\tilde{\chi}_1^\pm, \tilde{\chi}_1^0)$. In table 1 a summary of the SRs selections is presented.

| SR | L90 | L100 | L110 | L120 |
|---|---|---|---|---|
| Leading lepton $p_T$ [GeV] | $> 25$ | | | |
| $\Delta\phi_j$ [rad] | $> 1$ | | | |
| $\Delta\phi_b$ [rad] | $< 1.5$ | | | |
| $m_{T2}$ [GeV] | $> 90$ | $> 100$ | $> 110$ | $> 120$ |
| Leading Jet $p_T$ [GeV] | - | $> 100$ | $> 20$ | - |
| Second leading Jet $p_T$ [GeV] | - | $> 50$ | $> 20$ | - |

**Table 1.** Summary of signal regions used in the leptonic $m_{T2}$ analysis. The table is taken from Ref. [2].

## 4. Stop into stau in the dileptonic channel

In gauge-mediated supersymmetry breaking (GMSB) models the spin 3/2 partner of the graviton, called gravitino ($\tilde{G}$), can be the lightest supersymmetric particle. Moreover, in a generic R-parity conserving SUSY model like that, the scalar partners of charged leptons, $\tilde{l}_R$ and $\tilde{l}_L$, can mix to form two sleptons mass eigenstates. The lighter of the two sleptons eigenstates is denoted as $\tilde{\tau}_1$. In the hypothesis that staus are lighter than stops, the stop decay into stau is allowed. The analysis presented focuses on the decay chain in figure 2 with the $\tilde{t}$ decaying into a $\tilde{\tau}_1$ slepton and finally into gravitino [3].

After the preselection the events with same flavour leptons are required to have the invariant mass of the pair outside the 71 - 111 GeV range. The angle $\Delta\phi_b$ is required to be smaller than 1.5 and $\Delta\phi_j$ greater than 1.

Signal regions targeting high ($m_{\tilde{t}_1} > 250$ GeV) and low ($m_{\tilde{t}_1} < 250$ GeV) stop masses are than defined, using further selections on the number of jets, $m_{T2}^{ll}$ and $m_{T2}^{jl}$. The relevant backgrounds for this analysis are the same of the leptonic $m_{T2}$ analysis and are estimated using dedicated control regions for top and Z boson contributions.

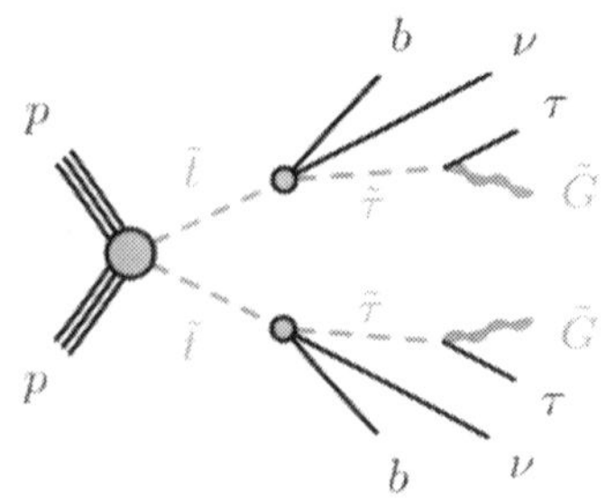


**Figure 2.** Top squark pair production diagram, decaying into $b\tilde{\tau}_1\nu$ followed by the $\tilde{\tau}_1$ decay into a $\tau$ lepton and the gravitino. $BR(\tilde{t}_1 \to b\tilde{\tau}_1\nu) = 1$ is assumed.

## 5. Summary of the results

A summary of the ATLAS searches for stop squark pair production using 20.3 fb$^{-1}$ LHC data collected by the ATLAS experiment during 2012 at 8 TeV energy have been presented. All the results are in agreement with the SM predictions and are interpreted in terms of simplified models where two stop squarks decay with 100% BR to a targeted channel.

A wide variety of analyses has been studied in two, one and zero leptons channels, summarized in figure 3. All the results are in agreement with the Standard Model and a large portion of the stop phase space has been excluded. A stop decay into $b + \tilde{\chi}_1^{\pm}$ with $m_{\tilde{t}_1}$ between 150 and 445 GeV is excluded at 95% CL in the case $m_{\tilde{\chi}_1^{\pm}} \sim m_{\tilde{t}_1}$ and $m_{\tilde{\chi}_1^0} = 1$ GeV. Stop with mass between 215 GeV and 530 GeV that decays into a top quark and a neutralino (red contour in figure 3) and between 90 and 170 GeV for the three body decay mode (purple contour in figure 3) are excluded at 95% CL.

The analysis described in section 3 contributes significantly to the whole stop exclusion scenario (light-blue areas in figure 3). The case $\tilde{t} \to b + \tilde{\chi}_1^{\pm}$ with $m_{\tilde{\chi}_1^{\pm}} = 2 \cdot m_{\tilde{\chi}_1^0}$ is excluded up to $m_{\tilde{t}} < 500$GeV, as well as the case $m_{\tilde{\chi}_1^{\pm}} = m_{\tilde{t}} - 10$GeV.

In the hypothesis of staus lighter than stops the exclusion limit at 95% CL in figure 4 has been obtained by the analysis described in section 4. A significant portion in the $(m_{\tilde{t}_1}, m_{\tilde{\tau}_1})$ plane is covered, for a stop and a stau lighter than 500GeV. The observed data are in agreement with the Standard Model background expectation, assuming $BR(\tilde{t}_1 \to b\tilde{\tau}_1\nu) = 1$ and $BR(\tilde{\tau}_1 \to \tilde{G}\tau) = 1$ in the exclusion limit calculation.

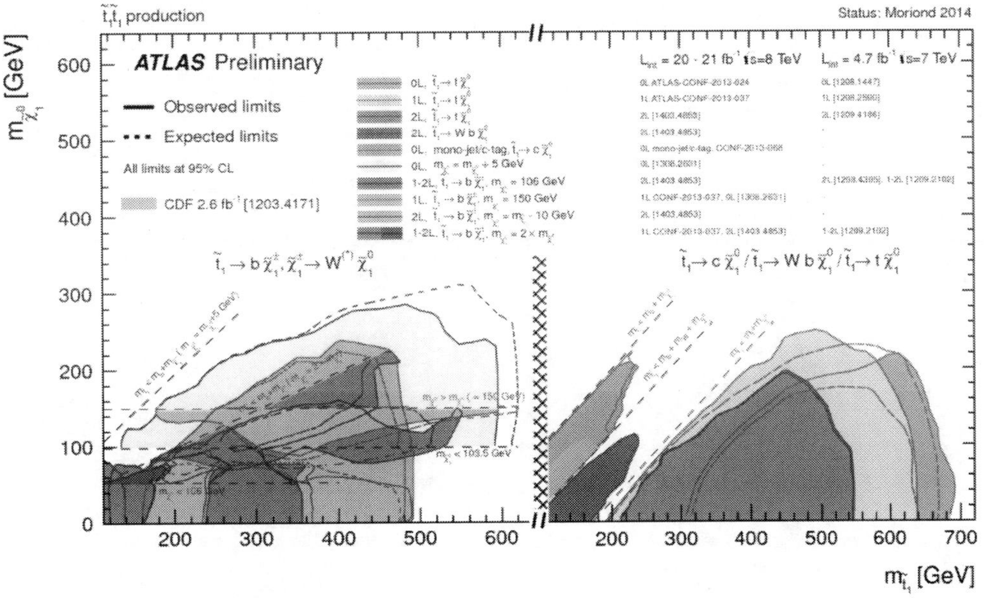


**Figure 3.** Summary of the ATLAS top squark pair production searches based on $20fb^{-1}$ of pp collision data taken at $\sqrt{s} = 8$ TeV and $4.7fb^{-1}$ of pp collision data taken at $\sqrt{s} = 7$ TeV. Exclusion limits at 95% CL in the stop - neutralino mass plane are shown. Image from [4].

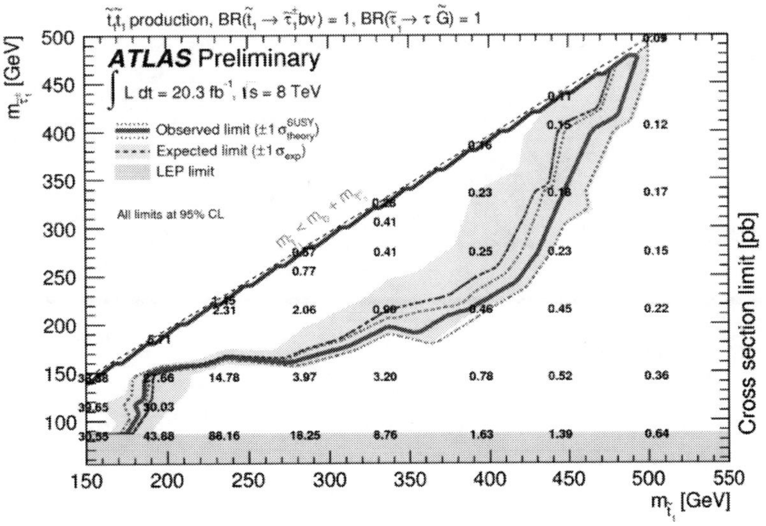


**Figure 4.** $\tilde{t}_1$ into $\tilde{\tau}_1$ exclusion limit at 95% CL using $20.3fb^{-1}$ of 8 TeV collision data, assuming $BR(\tilde{t}_1 \to b\tilde{\tau}_1\nu) = 1$, taken from [3]. The blue dashed line and the yellow band are the expected limit and its uncertainty. The thick solid line is the observed limit for the central value of the signal cross section. The dotted lines show the effect on the observed limit when varying the signal cross section by $\pm 1\sigma$ of the theoretical uncertainty. The green band represents the LEP limit on the $\tau$ slepton mass.

## References

[1] R. Barbieri and G. F. Giudice, Nucl. Phys. B **306** (1988) 63.
[2] ATLAS Collaboration, arXiv:1403.4853v1 [hep-ex], JHEP06 (2014) 124
[3] ATLAS Collaboration, ATLAS-CONF-2014-014
[4] ATLAS supersymmetry public results webpage (https://twiki.cern.ch/twiki/bin/view/AtlasPublic/SupersymmetryPublicResults)
[5] C. G. Lester and D. J. Summers, Phys. Lett. B463 (1999) 99–103, arXiv:hep-ph/9906349.
[6] A. Barr, C. Lester, and P. Stephens, J. Phys. G29 (2003) 2343–2363, arXiv:hep-ph/0304226
[7] ATLAS Collaboration, JINST 3, S08003 (2008)

5th Young Researcher Meeting, Trieste 2014 IOP Publishing
Journal of Physics: Conference Series **566** (2014) 012002 doi:10.1088/1742-6596/566/1/012002

# The LAGUNA/LBNO potential for Long Baseline neutrino physics

**Luca Agostino.** *On behalf of the LAGUNA-LBNO Consortium*
Laboratoire APC, Université Paris Diderot, 10 rue Alice Domon et Léonie Duquet 75013 Paris, FR

E-mail: luca.agostino@apc.univ-paris7.fr

**Abstract.** The LAGUNA/LBNO collaboration proposes a new generation neutrino experiment to address fundamental questions in particle and astroparticle physics. The experiment consists of a far detector, Liquid Argon (LAr) double phase Time TPC (Time Projection Chamber), the fiducial mass of the detector is set to 20 kt in its first stage. The detector will be situated at 2300 km from CERN: this long baseline provides a unique opportunity to study the neutrino flavour oscillations over the first and second oscillation maxima and to explore the L/E (Length over energy) behaviour. The near detector is based on a high-pressure argon gas TPC situated at CERN. I will detail the physics potential of this experiment for determining without ambiguity the mass hierarchy (MH) in its first stage and discovering CP violation (CPV) using the CERN SPS beam with a power of 750 kw. The impact of the assumptions on the knowledge of the oscillation parameters and the systematic errors are very important and will be shown in detail to prove the force of the experiment assuming realistic and conservative parameter values.

## 1. Introduction

The LBNO (Long Baseline Neutrino Observatory) project is the result of two European Design Studies: LAGUNA in 2008 and LAGUNA/LBNO in 2011. From the point of view of the neutrino physics, the main scientific goal is the study of long baseline neutrino oscillations. Important discoveries in the neutrino sector during the 2010-2012 period and in particular the precise measure of $sin^2 2\theta_{13}$ had an impact on the physics case of the experiment . In particular, as the neutrino mixing angle $\theta_{13}$ is big ($sin^2 2\theta_{13} = 0.084^{+0.005}_{-0.005}$ [3]), the Mass Hierarchy (MH) of neutrinos in first instance and, secondary, the CP violation discovery are now the most important items. A very long beam accelerated baseline from CERN to Pyhäsalmi (Finland), corresponding to a length of 2300 km, has been indicated as the first choice. The recommended technology for the far detector is the double phase LAr Large Electron Multiplier Time Projection Chamber (LAr LEM-TPC), known to provide excellent tracking and calorimetry performance installed very deep underground inside an existing mine. The main goals are the neutrino Mass Hierarchy determination (guaranteed at more than 5 $\sigma$) and the leptonic CP violation discovery, obtained by the analysis of the L/E dependence of the oscillation pattern and of both the $1^{st}$ and $2^{nd}$ oscillation maxima to distinguish the matter effect from the $\delta_{CP}$ effect. LBNO will be the first experiment to provide a whole 3-flavor neutrino oscillation analysis from the appearance and disappearance channels.

To realize the ambitious physics program in the most efficient way, the LBNO Consortium proposes an "incremental approach" involving two phases:

- Phase I - LBNO20: For a first phase, a 24 kt LAr active volume detector and the SPS beam from CERN providing 400 GeV proton energy and 750kW are considered. The LAr detector will be coupled to a magnetized iron calorimeter (MIND) with muon momentum and charge determination, that will collect an independent neutrino sample and will serve as a tail catcher for CERN beam events occurring in the LAr target. This first phase (4 years) will solve the MH problem with $\simeq$ 100% probability and will give some hints of the value of the $\delta_{CP}$ phase.
- Phase II - LBNO70 An increase of the detector mass up to 70 kt and a possible increase of the beam power up to 2 MW is envisaged to provide a very precise measure of the $\delta_{CP}$ assuring a coverage of the CP phase space of 75% at 3 $\sigma$ and 54% at 5 $\sigma$.

**Figure 1.** The baseline CERN to Pyhäsalmi.

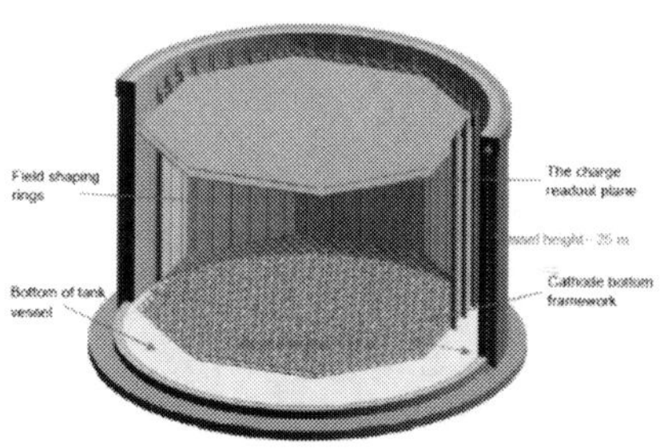


**Figure 2.** A sketch of the GLACIER type detector.

## 2. Mass Hierarchy determination

The first goal of the LBNO project is to determine the mass hierarchy ordering. It will be addressed during the first years of run. In this direction the aim of LBNO is to guarantee the discovery of the MH at about 100% power which means that the probability to give the wrong answer is nearly zero. To do that, we simulated 4 years of data taking with the SPS accelerator [6] at 750 kW using a 50% $\nu$ 50% $\bar{\nu}$ sharing, which is optimized for this measurement, and $4 \times 10^{20}$ Protons on Target (PoT). Results are shown in first instance in terms of the mean value $T_0$ of the test statistics $T$ defined as the difference between the $\chi^2$ of the event distribution evaluated assuming the normal (NH) and the inverted mass hierarchy (IH) as true.

$$T = \chi^2_{IH} - \chi^2_{NH} \tag{1}$$

where $\chi^2_{IH}$ ($\chi^2_{NH}$) is obtained by minimizing the $\chi^2$ defined as:

$$\chi^2 = \chi^2_{appear} + \chi^2_{disa} + \chi^2_{syst}. \tag{2}$$

Journal of Physics: Conference Series **566** (2014) 012002 doi:10.1088/1742-6596/566/1/012002

where

$$\chi^2_{appear} = 2\sum_{+/-}\sum_{E_\nu^{rec},p_T^{miss}} n_e(E_\nu^{rec}, p_T^{miss}; \mathbf{o}_{test}, \mathbf{f}_{test}) - n_e(E_\nu^{rec}, p_T^{miss}; \mathbf{o}_{true}, \mathbf{f}_{true}) + n_e(E_\nu^{rec}, p_T^{miss}; \mathbf{o}_{true}, \mathbf{f}_{true}) \ln \frac{n_e(E_\nu^{rec}, p_T^{miss}; \mathbf{o}_{true}, \mathbf{f}_{true})}{n_e(E_\nu^{rec}, p_T^{miss}; \mathbf{o}_{test}, \mathbf{f}_{test})}, \quad (3)$$

The number of the e-like events in a given $E_\nu^{rec} - p_T^{miss}$ bin is determined according to:

$$\begin{aligned} n_e(E_\nu^{rec}, p_T^{miss}; \mathbf{o}, \mathbf{f}) = {} & f_{sig} n_{e-sig}(E_\nu^{rec}, p_T^{miss}; \mathbf{o}) \\ & + f_{\nu_e} n_{\nu_e}(E_\nu^{rec}, p_T^{miss}; \mathbf{o}) + f_{\nu_\tau} n_{e,\nu_\tau}(E_\nu^{rec}, p_T^{miss}; \mathbf{o}) \\ & + f_{NC}(n_{NC\pi^0}(E_\nu^{rec}, p_T^{miss}; \mathbf{o}) + n_{mis-\nu_\mu}(E_\nu^{rec}, p_T^{miss}; \mathbf{o})), \end{aligned} \quad (4)$$

where $n_{e-sig}$, $n_{\nu_e}$, $n_{e,\nu_\tau}$, $n_{NC\pi^0}$, and $n_{mis-\nu_\mu}$ are the number of events for signal, intrinsic beam $\nu_e$, electrons from tau decay, neutral current, and mis-identified $\nu_\mu$, respectively.

The information from the disappearance channel is contained in the $\chi^2_{disa}$ term of total $\chi^2$ in Eq. 2, which is given by

$$\chi^2_{disa} = 2\sum_{+/-}\sum_{E_\nu^{rec}} n_\mu(E_\nu^{rec}; \mathbf{o}_{test}, \mathbf{f}_{test}) - n_\mu(E_\nu^{rec}; \mathbf{o}_{true}, \mathbf{f}_{true}) + n_\mu(E_\nu^{rec}; \mathbf{o}_{true}, \mathbf{f}_{true}) \ln \frac{n_\mu(E_\nu^{rec}; \mathbf{o}_{true}, \mathbf{f}_{true})}{n_\mu(E_\nu^{rec}; \mathbf{o}_{test}, \mathbf{f}_{test})}. \quad (5)$$

The number of $\mu$-like events is the sum of signal ($n_{\mu-sig}$) and $\tau \to \mu$ background ($n_{\mu,\nu_\tau}$) contributions and is calculated as

$$n_\mu(E_\nu^{rec}; \mathbf{o}, \mathbf{f}) = f_{sig} n_{\mu-sig}(E_\nu^{rec}; \mathbf{o}) + f_{\nu_\tau} n_{\mu,\nu_\tau}(E_\nu^{rec}; \mathbf{o}). \quad (6)$$

In Figures 3 and 4 $T_0$ is shown as a function of the true value of $\delta_{CP}$, for the assumed Normal and Inverted true mass hierarchy. It could be put in evidence the slight different strength of the two $T_0$ distributions due to the difference in term of oscillation probability between NH and IH. The value of $T_0$ is strongly dependent on the value of the mixing angle $\theta_{23}$. On the same Figures the effect of different assumptions of $\theta_{23}$ spanning some values between the first and the second octant solutions is shown. For an experiment like LBNO is then mandatory to ensure enough statistics to avoid the uncertainty due to this assumption. The retained value of $sin^2\theta_{23}$ for these studies is 0.44 which is in agreement with recent global fits [7], but does not correspond to the best signal-to-noise. The other sources of uncertainty, such as the error on $\theta_{23}$ or the density of the Earth, can only have a smaller impact.It is to be stressed here that, as it is evident from these Figures, the best and worst value of $T_0$ are reached for $\delta_{CP} = \frac{\pi}{2}, \frac{3}{2}\pi$.

The statistical power defined in details in [1] for the MH determination at a confidence level of $3\sigma$ or $5\sigma$ is shown in Figure 5 and 6 as a function of exposure.

For each confidence level the shaded area corresponds to the results assuming different values of $\delta_{CP}$ between 0 and $2\pi$. The extreme values are reached for $\delta_{CP} = \pi/2$ (left border) or $3\pi/2$ (right border), as has been explained above. This plot can be used to state the minimal exposure needed to get the desired confidence level with a certain probability. Usually experiments state their sensitivity with a probability to obtain the result of 50% (median sensitivity). LBNO can solve the mass hierarchy problem with a median sensitivity at $5\sigma$ C.L. within the first year of exposure (1.5e20 PoT). In addition one can see that LBNO has a probability of essentially 100% to discover the MH in either case for any value of $\delta_{CP}$. An exposure of slightly more that $2 \times 10^{20}$ PoT will guarantee that a $3\sigma$ C.L. is obtained, while a $5\sigma$ C.L. will be reached with less than $4 \times 10^{20}$ PoT, corresponding to about 4 years of SPS running.

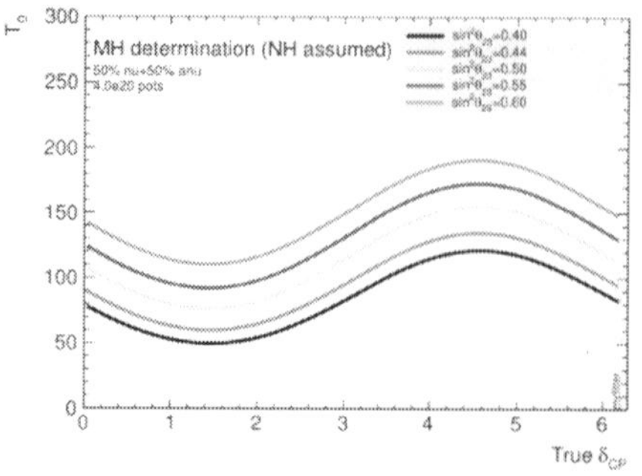


**Figure 3.** Mean value of the mass hierarchy test statistic for Normal Hierarchy as a function of true $\delta_{CP}$ for a total exposure of $4 \times 10^{20}$ PoT (or about 4 years of running at the SPS) and LBNO 20 kton detector.

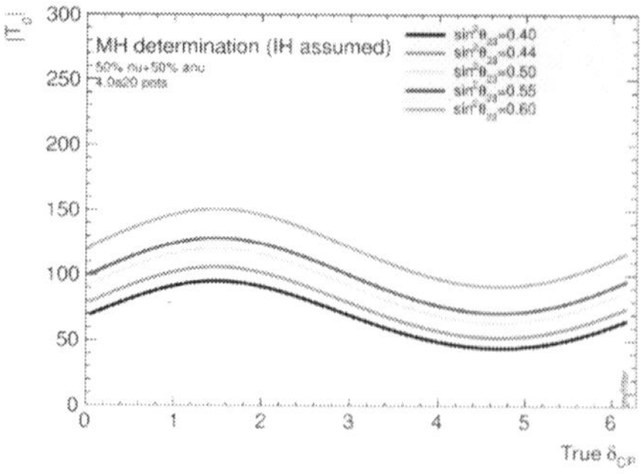


**Figure 4.** Mean value of the mass hierarchy test statistic for Inverted Hierarchy as a function of true $\delta_{CP}$ for a total exposure of $4 \times 10^{20}$ PoT (or about 4 years of running at the SPS) and LBNO 20 kton detector. Note the absolute value on y axis to directly compare with Figure 3.

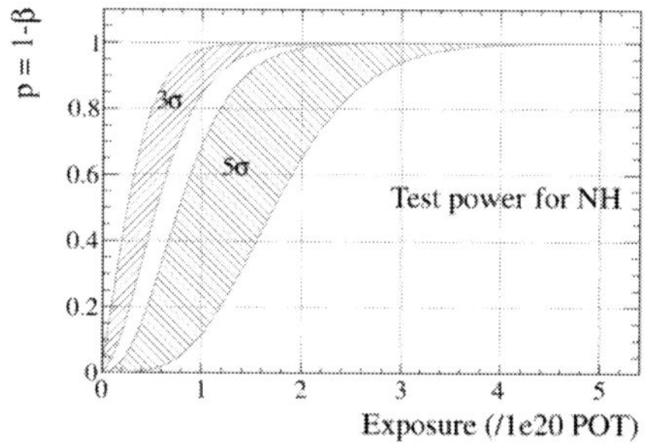


**Figure 5.** Statistical power as a function of exposure for the test of NH for $3\sigma$ and $5\sigma$ CL. The nominal central values for oscillation parameters have been assumed and the shaded bands correspond to the variation of $\delta_{CP}$.

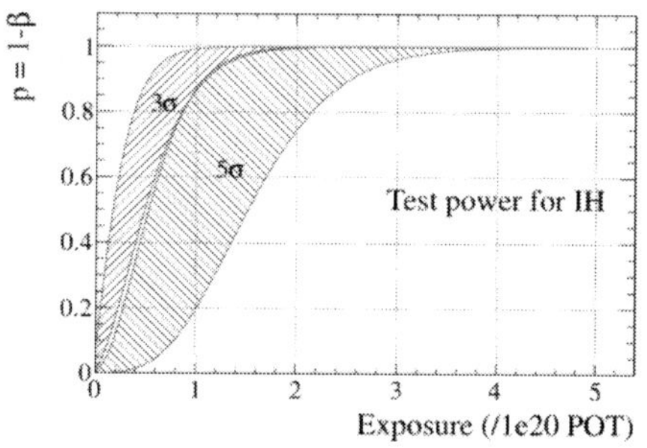


**Figure 6.** Statistical power as a function of exposure for the test of IH for $3\sigma$ and $5\sigma$ CL. The nominal central values for oscillation parameters have been assumed and the shaded bands correspond to the variation of $\delta_{CP}$.

## 3. CP violation

Once the ordering of the mass hierarchy has been determined, LBNO will continue to run and collect statistics in order to establish if the $\delta_{CP}$ phase is non vanishing and to measure it. The sensitivity to a non vanishing $\delta_{CP}$ is obtained in terms of tests statistics, testing the hypothesis of no-CP-violation evaluating the $\Delta\chi^2$ as a function of the value of $\delta_{CP}$. The evaluated $\Delta\chi^2$ is then plotted as a function of the value of $\delta_{CP}$ assumed as alternative hypothesis as in Figure 8. The coverage of the experiment is defined as the portion of the $\delta_{CP}$ phase space for which the value of $\Delta\chi^2$ is greater than 9 (at $3\sigma$ C.L.) or 25 (at $5\sigma$ C.L. ). To have a complete information about the LBNO sensitivity to discover the CP violation, the power for the $\delta_{CP}$ sensitivity is plotted following the same approach used for the MH. The statistical power of LBNO for CPV determination as a function of exposure is shown in Figure 7, for the two different $C.L.$s of 90%

and $3\sigma$ when a 20 kt detector fiducial mass is assumed. The bands limit curves correspond to the two most favorable cases, $\delta_{CP} = \pi/2$ or $3\pi/2$.

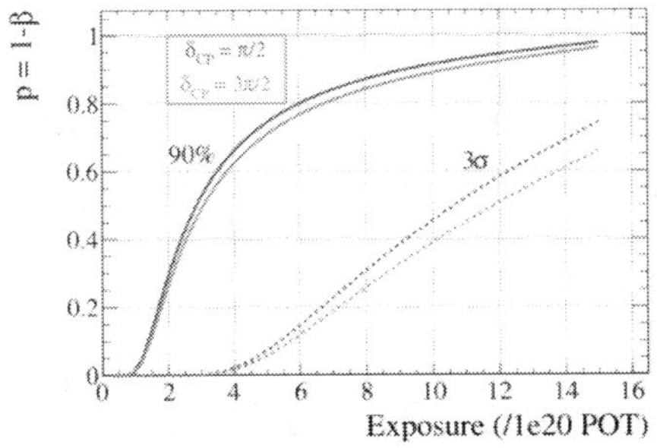


**Figure 7.** Statistical power for CPV discovery as a function of exposure for 90% and $3\sigma$ CL assuming NH. The far detector of 20 kton LAr and 750 kW SPS neutrino beam are assumed.

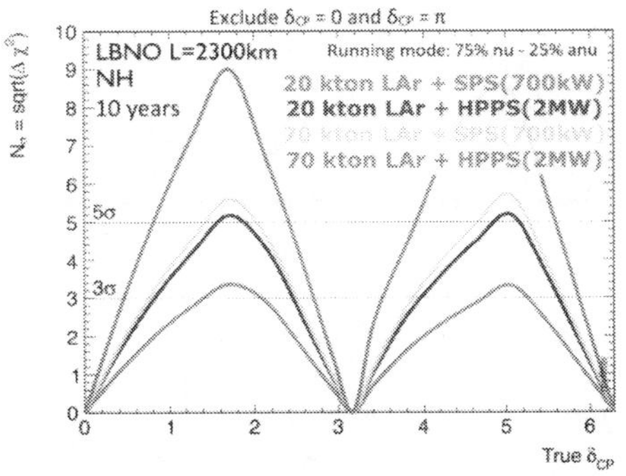


**Figure 8.** CPV sensitivity as a function of $\delta_{CP}$ for various upgrades of beam power with the HP-PS[8], and of the far detector mass, with 20 kton and 70 kton.

As detailed in previous paragraphs, LBNO Phase I can reach significant physics goals, in particular it is guaranteed to be fully conclusive for MH discovery with an expected $5\sigma$ C.L. over the full range of $\delta_{CP}$. On the other hand, the CPV sensitivity reach is more difficult to predict, since ultimately is dependent on the achievable systematic errors and on the true $\delta_{CP}$. The studied detailed so far have been realized using presently realistic errors on oscillation parameters and on the normalization of the signal and backgrounds. This is a extreme conservative assumption motivated by the aim to guarantee the success of the experiment independently of further future improvements. Of course, with the series of expected new measurements and possibly the addition of dedicated measurements from experiments on hadron-production and neutrino cross-sections, it is conceivable to think that the overall balance of errors could be reduced in the future, thereby improving further the expected CPV sensitivity of LBNO20.

In addition to these improvements, a complementary and/or alternative method to get better CP violation sensitivity is to increase the detector mass and the neutrino beam power. This will allow an important increase of the number of events and therefore a decrease of the statistical error around the 2nd oscillation maximum. Due to the natural cut-off of the muon-neutrino flux spectrum at low energy and the linear increase of the total neutrino cross-section with energy, the 2nd maximum is more difficult to study than the 1st one. But, as the second maximum is still accessible by LBNO given the very long baseline, a significant gain is obtained by populating this region with oscillation events.

This is one of the main goals of the LBNO70. The expected CPV sensitivity (in terms of $\Delta\chi^2$ as a function of $\delta_{CP}$ is shown in Figure 8 for various upgrades of beam power with the HP-PS, and of the far detector mass, from 20 kton to 70 kton. With a new powerful proton driver such as the conceptual HP-PS and a 70 kton detector mass, the coverage at $> 5\sigma$'s C.L. will be ~54% after 10 years.

Table 1 summarizes the coverages of the CP phase space for LBNO20 and LBNO70 considering the SPS and the HP-PS beams.

Journal of Physics: Conference Series **566** (2014) 012002 doi:10.1088/1742-6596/566/1/012002

| beam | 24 kt at 3 $\sigma$ | 70 kt at 3 $\sigma$ | 24 kt at 5 $\sigma$ | 70 kt at 5 $\sigma$ |
|---|---|---|---|---|
| SPS | 25% | 60% | / | 25% |
| HP-PS | 57% | 75% | 21% | 54% |

**Table 1.** Percentage coverage of the CP phase space for different detector masses with the SPS and the HP-PS beam.

## 4. Conclusions

The next generation neutrino detectors will address fundamental questions in particle and astroparticle physics. They will determine the Mass Hierarchy of Neutrinos, this will open the way to the CP symmetry violation discovery and, if CP is violated, to the precise measure of the $\delta_{CP}$ phase. A complete astrophysics program will be hopefully addressed by measuring the Supernovae background Neutrino and Neutrinos from Core Collapse Supernovae and performing at the same time precise measurements of atmospheric and solar neutrinos. New limits on proton lifetime will be fixed.

In this context, the LBNO experiment design is capable to meet the fore-mentioned scientific goals. It is the result of two dedicated Design Studies: LAGUNA and LAGUNA/LBNO, which proposed an incremental phased approach considering a 24 kt (LBNO20) upgradable to 70 kt (LBNO70) double phase Liquid Argon GLACIER-type detector to be installed inside the Pyhäsalmi mine in Finland at 2300 km from CERN where a dedicated neutrino beam will be produced.

Beam based experiments are the only capable to run in both $\nu$ and $\bar{\nu}$ configurations by changing the horn polarity allowing a full measure of the $3\nu$ flavor paradigm. In addition, the deep underground location of such detector is ideal for astrophysics measurements and proton decay search. LBNO is a proposed long baseline experiment with a complete design study including a detailed and motivated physics reach, realistic simulations and layout concepts of the neutrino beam, underground construction plan, costing estimations and risk assessments. The detector response as well as the analysis software development and the engineer challenges will be the main goals of the CERN experiment WA105, the LBNO prototype (LBNO-DEMO) [9] .

## References


[1] LAGUNA-LBNO Collaboration "The mass-hierarchy and CP-violation discovery reach of the LBNO long-baseline neutrino experiment" JHEP05(2014)094 arXiv:1312.6520v3 [hep-ph]

[2] F. P. An *et al.* [DAYA-BAY Collaboration], "Observation of electron-antineutrino disappearance at Daya Bay," Phys. Rev. Lett. **108**, 171803 (2012) [arXiv:1203.1669 [hep-ex]].

[3] Daya Bay collaboration PRL 112, 061801 (2014)

[4] L. Agostino *et al.* [MEMPHYS Collaboration] "Study of the performance of a large scale water-Cherenkov detector (MEMPHYS)" JCAP **1301** (2013) 024 [arXiv:1206.6665[hep-ex]].

[5] A. Stahl *et al.*, Expression of Interest for a "Very long baseline neutrino oscillation experiment (LBNO)" submitted to CERN SPSC, June, 2012. (CERN-SPSC-2012-021 (SPSC-EOI-007))

[6] E. Shaposhnikova *et al.*,"Recent intensity increase in the CERN accelerator chain", CERN-AB-2005-029.

[7] G. L. Fogli, E. Lisi, A. Marrone, D. Montanino, A. Palazzo and A. M. Rotunno, "Global analysis of neutrino masses, mixings and phases: entering the era of leptonic CP violation searches" Phys. Rev. D **86** (2012) 013012 [arXiv:1205.5254 [hep-ph]].

[8] Y. Papaphilippou et al., "Design options of a High-Power Proton Synchrotron for LAGUNA-LBNO", THPWO081, IPAC13 proceedings.

[9] The LBNO-DEMO Collaboration, "Technical Design Report for a large neutrino detectors prototyping and phased performance assessment in view of a long-baseline oscillation experiment"

Journal of Physics: Conference Series **566** (2014) 012003 doi:10.1088/1742-6596/566/1/012003

# Neutrino mass scale in the era of precision cosmology

**M. Gerbino**[1]

[1] Physics Dep. and INFN, Università di Roma "Sapienza", P.le A. Moro, 2, 00185, Rome, Italy

E-mail: `martina.gerbino@roma1.infn.it`

**Abstract.** In this work, we study how recent cosmological datasets in combination with particle physics results are able to constrain the neutrino mass scale. In particular, we present current bounds on the electron neutrino mass $m_\beta$, the effective Majorana mass $m_{\beta\beta}$ and the total neutrino mass $\Sigma m_\nu$ and we discuss the sensitivity required to future experiments in order to address the issue of neutrino hierarchy.

## 1. Introduction

Neutrinos are weakly interacting and neutral particles. Because of their elusive nature, many neutrino properties are still debated. In particular, it is not clear if neutrinos are Dirac or Majorana particles. In the former case, neutrino and antineutrino are separate particles and this assures the lepton number to be preserved in interactions; in the latter case, neutrino is its own antiparticle. What is not debated anymore is the fact that neutrinos are massive. The major evidence comes from oscillation experiments, measuring transitions in flight between neutrino flavors: each neutrino flavor is a superposition of mass eigenstates with coefficients given by the neutrino mixing matrix. The latter is parameterized via neutrino mixing angles $\theta_{ij}$ and phases, one $\delta$ if neutrinos are Dirac particles, responsible for CP violation, two additional phases $\phi_1$ and $\phi_2$ if neutrinos are Majorana. In this work, we assume $\delta$ to be negligible, without any loss of generality.

Oscillation experiments can help to constrain these mixing parameters (see [1] for an updated review). In particular, they are sensitive to the mixing angles $\theta_{ij}$ and to the squared mass differences: $\delta m^2 \approx 10^{-5}\,\mathrm{eV}^2$, the difference between the first two eigenstates, and $\Delta m^2 \approx 10^{-3}\,\mathrm{eV}^2$, the difference between the third eigenstate and the mean value of the first two. $\Delta m^2$ can be greater or lower than zero, identifying two mass hierarchies: in the former case, we have normal hierarchy (NH), with the third eigenstate as the most massive; in the latter case, the third eigenstate is the lightest one and the hierarchy is inverted (IH). However, oscillation experiments are not sensitive to the sign of $\Delta m^2$, so they cannot distinguish between the two hierarchies. Moreover, they cannot establish the absolute mass scale of neutrinos, which has to be probed by other kind of observables.

As far as particle physics is concerned, the main experiments looking for the absolute mass scale involve beta and double-beta decay. The goal of beta decay experiments is to assess the shape of the decay spectrum close to the endpoint, since it shows a different behavior as a function of neutrino mass $m_\beta$. Future experiments aim to reach ten times the current sensitivity ($\approx 2\,\mathrm{eV}$ [2]). This is for example the experimental goal of Katrin [3]. While less sensitive than other observables, beta decay is a very powerful probe since it is completely model independent.

Journal of Physics: Conference Series **566** (2014) 012003 doi:10.1088/1742-6596/566/1/012003

Double-beta decay involves the emission of two electrons and two antineutrinos. If neutrinos were Majorana particles, than the two antineutrinos could annihilate and electrons would be the only decay products, giving rise to a process known as neutrinoless double-beta decay. Double-beta decay experiments put limits on the half life $T_{1/2}$ of the isotope involved in the neutrinoless decay. Theoretically, $T_{1/2}$ is expressed as a function of the phase space factor, which is easy to calculate, the nuclear matrix element (NME), more difficult to assess, and the Majorana effective mass $m_{\beta\beta}$ (a combination of massive eigenstates weighted by mixing angles), which is the quantity of interest from the point of view of particle physics. Taking into account the different possible values for NME from nuclear models, the allowed range for the upper limit of $m_{\beta\beta}$ is $[0.2 - 0.4]$ eV [4, 5]. As a result, the main hurdle to deal with when considering double-beta decay experiments is the great uncertainty arising from NME calculation; however, the power of double-beta decay experiments lays in the fact that evidence for $m_{\beta\beta} \neq 0$ would definitively solve the puzzle about neutrino nature [6].

We will now discuss what kind of information we can get from cosmological observables: the cosmic microwave background (CMB) and the large scale structures (LSS), both of them being sensitive to the sum of neutrino masses $\Sigma m_\nu$ [8]. As far as the CMB is concerned, massive neutrinos contribute to the total energy density of the Universe, affecting the expansion rate. The main observable effect is the shift and smearing of the peaks in the CMB power spectrum (the harmonic transform of the two-point correlation function of the anisotropy temperature map). Massive neutrinos also affect the way cosmological structures grow, since they prevent them to form below a certain scale, the neutrino free-streaming scale. The main effect is visible in the matter power spectrum (Fourier transform of the two-point correlation function of the matter density distribution): more massive neutrinos determine less power on small scales. An additional probe comes from the baryon acoustic oscillations (BAO), visibile as wiggles in the matter power spectrum. BAO are the results of acoustic oscillations undergone by the primordial plasma, due to the counterbalance between radiative pressure and gravitational attraction. Since neutrinos are part of the primordial content of the Universe, different neutrino masses would leave their imprint in the BAO. The current cosmological limits on the total mass are the tightest constraints on the neutrino mass scale. However, the price to pay is that they are highly model dependent.

## 2. Data analysis

We choose to follow a Bayesian approach and perform a MCMC analysis. The baseline model is given by the recent global fit to the oscillation parameters from [7]. We assume $\delta m^2$, $\Delta m^2$, $\sin\theta_{12}$, $\sin\theta_{13}$ to be Gaussian distributed around their best fit value, with variance given by the semi-width of the $1\sigma$ range. The parameter vector is then $\delta m^2$, $\Delta m^2$, $\sin\theta_{12}$, $\sin\theta_{13}$, $\phi_1$, $\phi_2$, $\Sigma m_\nu$, the last one in order to fix the scale mass. We assume flat priors and build our likelihood as the product of the single Gaussian likelihoods.

We then add further datasets to the baseline. This information are added via additional Gaussian likelihood with parameters reported below (see [9] for further details):

- **Gerda**: $T_{1/2} > 2.1 \cdot 10^{25}$ yr, Gerda experiment, phase 1 [4];
- **P+WP+high$\ell$**: $\Sigma m_\nu < 0.33$ eV , Planck [10], ACT [11] and SPT [12] TT power spectra plus WMAP9 low$\ell$ polarization (P+WP);
- **CMB+LSS1a**: $\Sigma m_\nu = (0.36 \pm 0.10)$ eV, P+WP marginalized over the weak lensing amplitude parameter $A_L$, CMASS measurements from [13], CFHTlens [14], GGlensing [15] and BAO (6dFGS and LOWZ) [16];
- **CMB+LSS2**: $\Sigma m_\nu = (0.38 \pm 0.11)$ eV, same as CMB+LSS1a, replacing CMASS measurements from [13] with [17];

- **CMB+LSS3**: $\Sigma m_\nu = (0.324 \pm 0.099)\,\mathrm{eV}$, same as CMB+LSS1a, replacing CMASS measurements from [13] with [18];
- **CMB+LSS4**: $\Sigma m_\nu = (0.27 \pm 0.11)\,\mathrm{eV}$, same as CMB+LSS1a, replacing CMASS measurements from [13] with the BAO only contraints from [19];
- **CMB+LSS1b**: $\Sigma m_\nu = (0.35 \pm 0.10)\,\mathrm{eV}$, same as CMB+LSS1a, replacing marginalized P+WP with WMAP9 [20];
- **CMB+LSS1c**: $\Sigma m_\nu = (0.27 \pm 0.12)\,\mathrm{eV}$, same as CMB+LSS1a, replacing marginalized P+WP with non-marginalized P+WP.

Few additional words have to be spent for the Gerda case. We choose to assess the impact of NME uncertainty introducing the nuisance parameter $\xi$ and marginalizing over it [21]. $\xi$ is defined as the ratio between a reference value for NME and its unknown exact value. We write down $T_{1/2}$ as a function of $m_{\beta\beta}$ and $\xi$ and define the likelihood of $T_{1/2}$ given $m_{\beta\beta}$ and $\xi$. This likelihood is chosen to be a semi-Gaussian in $m_{\beta\beta}$, centered in $m_{\beta\beta} = 0$, with variance given by the lower limit on $T_{1/2}$ from Gerda experiment. The prior on $\xi$ is flat since it represents the less informative choice, with $\xi$ spanning the range $[0.5 - 2]$, which is twice the uncertainty on the NME value.

## 3. Results

Fig.1 represents the parameter space explored by the chains of the MCMC analysis for NH (red) and IH (blue) for the baseline model. We can identify the physical lower limit for the mass scale

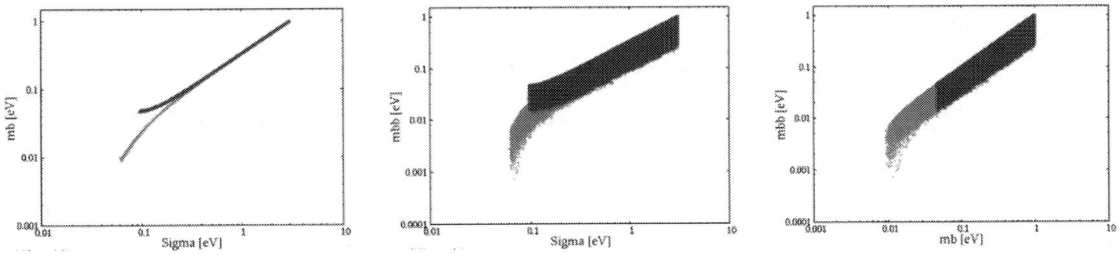


**Figure 1.** Portion of the available parameter space for the baseline model effectively explored in the planes $\Sigma m_\nu - m_\beta$, $\Sigma m_\nu - m_{\beta\beta}$, $m_\beta - m_{\beta\beta}$. $\Sigma m_\nu$ is indicated here as "Sigma".

parameters as a result of oscillation constraints and the degenerate region, at higher masses, where the two hierarchies are indistinguishable. From Fig.1, it is clear that particle physics experiments have to reach a sensitivity lower than 0.1 eV in order to distinguish between the two hierachies. In Fig.2, we can see what happens if we add information from Gerda, CMB and LSS.

First of all, notice the different sensitivity among datasets. Gerda provides the broadest constraints, while the tightest ones come from the cosmological datasets. Notice also the different location of the peaks for the two cosmological datasets. This is due to the different constraints on $\Sigma m_\nu$ coming from the two datasets, which drive the constraints on the other parameters. We can briefly comment about the impact of NME marginalization. In Fig.3, we compare the case of $\xi$ as a free parameter and $\xi = 1$. It is clear that the main impact of this nuisance parameter is at higher masses, making them more probable.

What can we learn from the other datasets? In Fig.4, the one-dimensional posterior probabilities for the mass scale parameters are reported. As previously said, the different location of the peak of the $\Sigma m_\nu$ posteriors coming from the various CMB+LSS analysis [9], also have an impact on

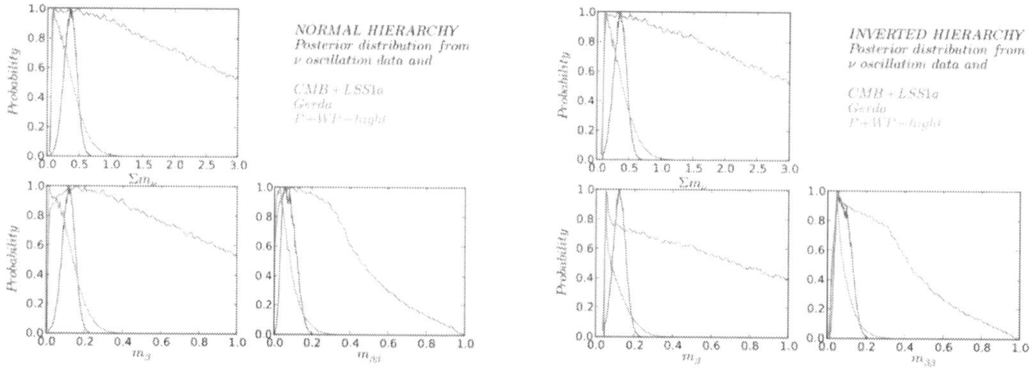


**Figure 2.** One-dimensional posterior probability of the mass scale parameters for the listed datasets. NH (IH) on the left (right).

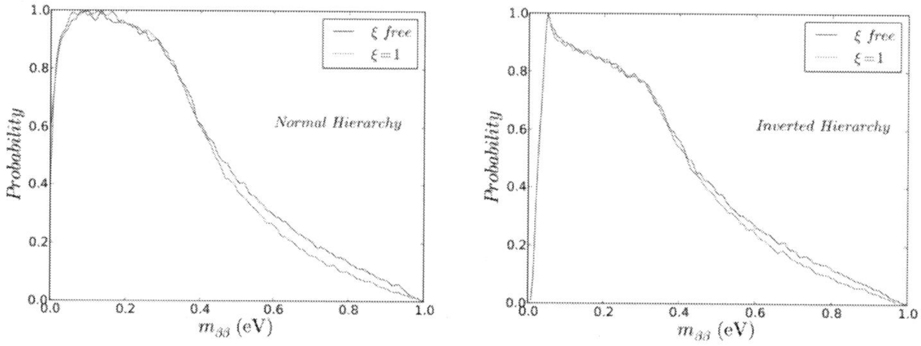


**Figure 3.** Impact of marginalization over $\xi$ nuisance parameter on the $m_{\beta\beta}$ posterior probability for the Baseline+Gerda combination.

the probability distributions of the other mass-scale parameters, highlighting how effective the combination of cosmological datasets and particle physics experiments could be.
The lower limit is still the effect of oscillation data, but all the cosmological datasets hint for $m_\beta$ and $m_{\beta\beta}$ to be in the region $[0-0.2]$ eV. We want to recall that the expected sensitivity from Katrin is $m_\beta < 0.2$ eV and the expected sensitivity from next generation double-beta experiments is $m_{\beta\beta} < 0.1$ eV. As a result, it appears that future experiments would not be able to assess definitive observations. In particular, to be able to distinguish between the two hierarchies, a sensitivity of order of 0.05 eV is required, as shown by the present analysis. This is evident from the left panel in Fig.5, where the $1\sigma$ and $2\sigma$ ranges for $m_{\beta\beta}$ for the different combinations of datasets are reported. The vertical dashed line is the future expected sensitivity. Obviously, increasing the sensitivity on $\Sigma m_\nu$ from cosmology would result in better constraints on $m_{\beta\beta}$ and $m_\beta$. In addition, the right plot in Fig.5 shows the $2\sigma$ ranges for $m_{\beta\beta}$ assuming the constraint on $\Sigma m_\nu$ from the dataset CMB+LSS1a. We can see that reducing the variance by a factor of two (center) and ten (bottom) with respect to the current one (top) would almost halve the uncertainty on $m_{\beta\beta}$.

Journal of Physics: Conference Series **566** (2014) 012003 doi:10.1088/1742-6596/566/1/012003

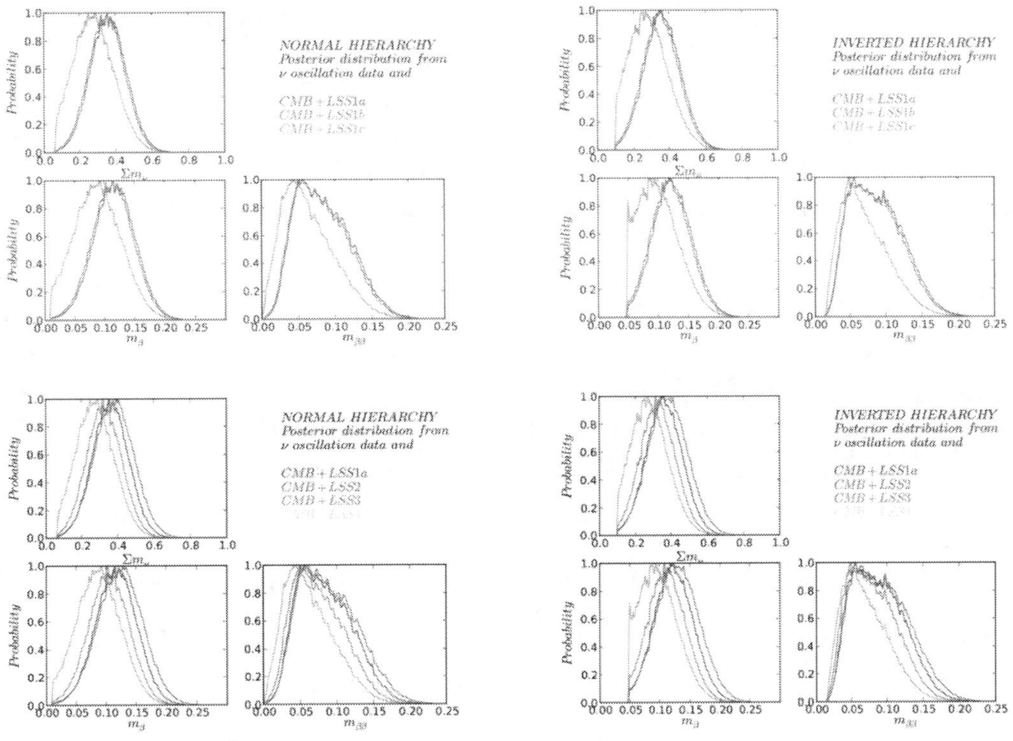


**Figure 4.** One-dimensional posterior probability distributions of the mass scale parameters for the listed datasets. NH (IH) on the left (right).

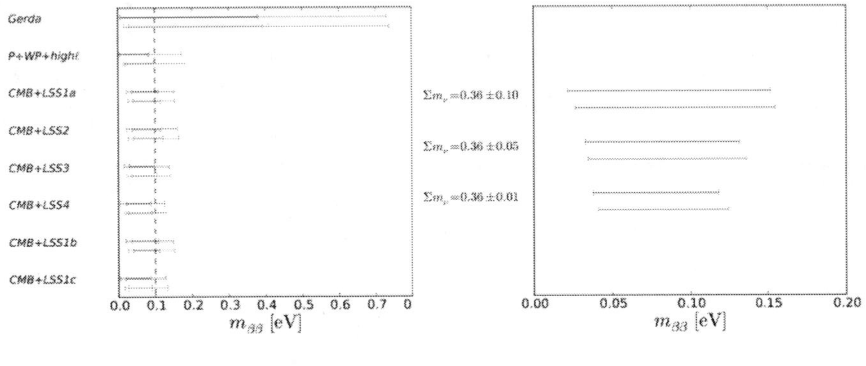


**Figure 5.** Left panel: $1\sigma$ (darker) and $2\sigma$ (lighter) ranges of $m_{\beta\beta}$ for the indicated datasets. In blue (red), constraints for normal (inverted) hierarchy. The vertical line is the future double-beta decay experiment sensitivity. Right panels: expected limits on $m_{\beta\beta}$ as a function of current and forecasted limits on $\Sigma m_\nu$ (in eV).

## 4. Conclusions

In this work, we have shown how a combination of cosmological and particle physics datasets is able to constrain the neutrino mass scale. In particular, we have shown that combining oscillation results with CMB and LSS data yields the tightest constraints on the effective electron neutrino mass, the Majorana mass and the total neutrino mass. Moreover, improving the observational bounds on $\Sigma m_\nu$ would result in stronger constraints on $m_{\beta\beta}$. Finally, it has also been shown that, given the current bounds on $\Sigma m_\nu$ from cosmology, a sensitivity of better than 0.1 eV is required in order to discriminate between normal and inverted hierarchy.

**Acknowledgments**
The author would like to thank Massimiliano Lattanzi and Alessandro Melchiorri for useful discussions.

## References

[1] Olive K A *et al.* (Particle Data Group), *Chin. Phys. C*, 38, 090001 (2014)
[2] Aseev V N *et al.* 2011, *Phys. Rev. D* **84** 112003; Weinheimer C *et al.* 1999, *Phys. Lett.* **B460** 219
[3] `http://www.katrin.kit.edu`
[4] Agostini M *et al.* [GERDA collaboration], *Phys. Rev. Lett.* **111** 122503
[5] Bhupal Dev P S , Goswami S, Mitra M and Rodejohann W 2013, *Phys. Rev. D* **88** 091301
[6] Schechter J and Valle J W F, *Phys. Rev. D* **25** 2951
[7] Forero D V , Tortola M , Valle J W F , arXiv:1405.7540
[8] Lesgourgues J and Pastor S, *AdHEP* **2012**, (2012)
[9] Beutler F *et al.* [BOSS Collaboration], *Mon. Not. Roy. Astron. Soc.* **444**, 3501 (2014), arXiv:1403.4599 [astro-ph.CO].
[10] Ade P A R *et al.* [Planck Collaboration], *Astron. Astrophys.* (2014), arXiv:1303.5076 [astro-ph.CO].
[11] Sievers J L, Hlozek R A, Nolta M R, Acquaviva V, Addison G E, Ade P A R, Aguirre P and Amiri M *et al.*, *JCAP* **1310**, 060 (2013), arXiv:1301.0824 [astro-ph.CO]
[12] Hou Z, Reichardt C L, Story K T, Follin B, Keisler R, Aird K A, Benson B A and Bleem L E *et al.*, *ApJ* **782**, 74 (2014), arXiv:1212.6267 [astro-ph.CO].
[13] Beutler F *et al.*, *Mon. Not. Roy. Astron. Soc.* **443**, 1065 (2014), arXiv:1312.4611 [astro-ph.CO].
[14] Kilbinger M *et al.*, *Mon. Not. Roy. Astron. Soc.* **430**, 2200 (2013)
[15] Mandelbaum R, Slosar A, Baldauf T, Seljak U, Hirata C M, Nakajima R, Reyes R and Smith R E, *Mon. Not. Roy. Astron. Soc.* **432**, 1544 (2013), arXiv:1207.1120 [astro-ph.CO].
[16] Beutler F. et al., *Mon. Not. Roy. Astron. Soc.* **416**, 3017 (2011); Anderson L *et al.*, *Mon. Not. Roy. Astron. Soc.* **441**, 24 (2014), arXiv:1312.4877 [astro-ph.CO]; Tojeiro R *et al.*, *Mon. Not. Roy. Astron. Soc.* **440**, 2222 (2014), arXiv:1401.1768 [astro-ph.CO].
[17] Chuang C H *et al.*, arXiv:1312.4889 [astro-ph.CO].
[18] Samushia L *et al.*, *Mon. Not. Roy. Astron. Soc.* **439**, 3504 (2014), arXiv:1312.4899 [astro-ph.CO].
[19] Anderson L *et al.*, *Mon. Not. Roy. Astron. Soc.* **441**, 24 (2014), arXiv:1312.4877 [astro-ph.CO].
[20] Bennett C L, Larson D, Weiland J L, Jarosik N, Hinshaw G, Odegard N, Smith K M and Hill R S *et al.*, *ApJS* **208**, 2 (2013).
[21] Minakata H, Nunokawa H, Quiroga A A, arXiv:1402.6014

Journal of Physics: Conference Series **566** (2014) 012004 doi:10.1088/1742-6596/566/1/012004

# Cosmology on the largest scales with intensity mapping

**Stefano Camera[1], Mário G Santos[2,3,1], Pedro G. Ferreira[4] and Roy Maartens[2,5]**

[1] CENTRA, Instituto Superior Técnico, Universidade de Lisboa, Avenida Rovisco Pais 1, 1049-001 Lisboa, Portugal

[2] Department of Physics, University of the Western Cape, Bellville 7535, South Africa

[3] SKA SA, 3rd Floor, The Park, Park Road, Pinelands, 7405, South Africa

[4] Astrophysics, University of Oxford, Dennis Wilkinson Building, Keble Road, Oxford OX1 3RH, UK

[5] Institute of Cosmology & Gravitation, University of Portsmouth, Dennis Sciama Building, Portsmouth P01 3FX, UK

E-mail: `stefano.camera@tecnico.ulisboa.pt`

**Abstract.** We review the state of the art of the study of the cosmic structure on ultra-large scales as is forecast to be achievable by the oncoming generation of intensity mapping experiments. We focus on intensity maps of the redshifted 21 cm line radiation of neutral hydrogen (HI) in the post-reionisation Universe. Such measurements will be performed by future radio telescopes such as for instance the Square Kilometre Array and will allow for surveying the biggest volume ever of cosmic structure. After having shown why it is valuable to scrutinise such extremely large cosmic scales – they will supply crucial information about the physical processes at play at early times – we concentrate on primordial non-Gaussianity as a working example. We illustrate that HI intensity mapping experiments can place tight bounds on different inflationary scenarios via constraining the non-Gaussianity parameter, $f_{\rm NL}$, with an error close to 1.

## 1. Introduction

The study of the largest cosmological scales and of the peculiar phenomena there occurring has recently seen a renewed interest. This has been mostly driven by the forthcoming generation of experiments which will scrutinise the cosmic large-scale structure with unprecedented accuracy over the biggest volume ever surveyed. However, if on the one hand such a new, unexplored landscape is a rich source of information valuable to strengthen our understanding of the Universe, on the other hand it is utterly difficult, technically speaking, to access those extremely large scales.

Following References [1, 2], we here review the state of the art concerning the effort put for reaching those ultra-large scales as well as the most important achievable scientific outcomes. More specifically, we shall briefly present the most interesting effects which may occur on the largest cosmic scales, such as deviations from the Newtonian prediction which can either further confirm general relativity or hint at a modified behaviour of gravity, or non-standard halo clustering due to a possible amount of non-Gaussianity in the statistical distribution of primordial density fluctuations. Hence, we shall focus on primordial non-Gaussianity and show what we could attain with one of the envisaged techniques, namely the so-called intensity mapping.

Surveying the large-scale cosmic structure by means of intensity maps of a particular line, such as for instance the redshifted 21 cm line of neutral hydrogen (HI), is a newly proposed technique (see [3] and references therein). It is still in its infancy but promises to obtain great results, being by the way pretty cheap compared to usual galaxy surveys. Instead of making a catalogue of HI emitting galaxies as detected above a certain flux threshold, an intensity mapping experiment effectively makes a map of large-scale fluctuations in 21 cm temperature by measuring the integrated emission in one big pixel. By doing so, it bypasses the need of resolving individual galaxies, thus being much similar to a map of the cosmic microwave background (CMB) radiation. Thanks to the high frequency (and thus redshift) resolution of oncoming radio telescopes, this intensity mapping methodology renders it possible to survey efficiently extremely large volumes.

## 2. Ultra-large scale effects

The investigation of the cosmic structure on the largest scales presents many advantages and can greatly improve our confidence in the theoretical model we employ to understand the Universe and its evolution. For a start, the modes related to the largest separations are well described by linear theory. Therefore, we do not have to rely on approximations coming from numerical simulations to take non-linear effects properly into account. This is a major gain, for numerical simulations are time consuming and they always imply a certain degree of ad hoc implementation [4, 5]. Closely related to this is the fact that on smaller scales, not only the growth of perturbations becomes non-linear, but also baryonic effects begin to play a more and more relevant rôle. Unfortunately though, the interplay of non-gravitational physical processes occurring on small scales – energy exchanges, radiative cooling, gas accretion to name a few – is very resilient to theoretical modelling (see e.g. [6]).

Having now clear in mind the pros of pursuing cosmology on ultra-large scale, let us briefly describe two of the most important effects that will be in principle detectable, namely general relativistic corrections to cosmological observables and primordial non-Gaussianity.

### *2.1. Testing gravity on the largest scales*

The vast majority of the predictions hitherto employed to analyse experimental data coming from surveys of the large-scale structure have relied on Newtonian or quasi-Newtonian approximations. This has proven to be fairly adequate for past and present surveys, since their interest is galaxy clustering well below the horizon scale. Often, so-called redshift-space distortions are also included in the analysis, and the contribution of weak lensing convergence and magnification to matter over-densities are sometimes considered too. However, there are much more effects, usually subdominant with respect to number density fluctuations and redshift-space distortions, but whose magnitude tends to increase as we go to larger scales. Specifically, the full relativistic analysis includes terms that are suppressed on sub-horizon scales, viz. velocity (or Doppler) terms, Sachs-Wolfe and integrated Sachs-Wolfe type terms, and time-delay contributions [7–9]. One the one hand, this implies that future wide and deep surveys will need to employ an accurate, precise modelling accounting not only for redshift-space distortions and weak lensing effects, but for all geometric and relativistic corrections. On the other hand, this also means that a clear detection of such relativistic corrections will signify a further confirmation of the goodness of Einstein's general relativity – even on scales farthest from those where it has been tested so far.

However, the largest cosmic scales might as well represent the very ground where Einsteinian predictions fail. Indeed, the last decade has seen a great endeavour by the cosmologists' community in the search for explanations of the observed late-time cosmic acceleration alternative to the standard cosmological constant hypothesis. The main idea inspiring such an effort is that behind the Universe's accelerated expansion there may be not some exotic component – for instance dark energy – which we never directly detected but nonetheless accounts for more than 70% of the total energy budget, bur rather a modification to the gravity law occurring on cosmological scales. (We refer the reader to Clifton et al. [10] for an exhaustive review on cosmology in modified gravity theories.) Since the accelerated expansion of the cosmos is a late-time phenomenon, it is natural to look for hints concerning the physical nature of the underlying mechanism on very large scales. Clearly, the detection of a modified gravity signature will be such an enormous breakthrough that it appears worth scrutinising the ultra-large scales.

### 2.2. *Primordial non-Gaussianity*

The standard inflationary scenario predicts almost Gaussian primordial fluctuations which successively grow by gravitational accretion to form the large-scale structure we observe today (e.g. [11]). However, inflation is rather a paradigm than an established theory and at the present day there is no such 'standard model' universally agreed upon. Quite on the contrary, many viable scenarios have been proposed, and a deeper understanding of the mechanism which drove inflation represents one of the cardinal goals of contemporary cosmology. To this aim, the study of primordial non-Gaussianity stands as a major tool at our disposal. Indeed, if the distribution of primordial density perturbations is not Gaussian, it cannot be fully described by a power spectrum; we rather need higher-order moments such as the bispectrum. In particular, different models of inflation give rise to different bispectrum shapes. Thence, it is in principle possible to discriminate amongst different inflationary models by measuring a particular shape of non-Gaussianity.

For instance, standard single-field inflation generates negligibly small deviations from Gaussianity. They are called of the local shape, where one of the three bispectrum wavenumbers has got a much smaller magnitude than the other two. Local-shape non-Gaussianity can also be generated when an additional light scalar field other than the inflaton contributes to the observed curvature perturbations, e.g. in curvaton models [12] or in multi-field models [13]. Besides, there are inflationary models with a non-standard inflaton kinetic term which contains higher-order derivatives of the field itself [14]. Their primordial bispectrum is maximised for configurations where the three wavevectors have approximately the same amplitude and falls under the name of equilateral-type primordial non-Gaussianity [15]. Eventually, we have also so-called squashed configurations [16] or shapes of the bispectrum which are nearly orthogonal to both the local and equilateral forms [17].

The method par excellence hitherto employed to constrain primordial non-Gaussianity has relied on measuring the CMB temperature anisotropy bispectrum [18]. However, primordial non-Gaussianity also induces an additional scale and redshift dependence in a biased tracer of the underlying matter distribution [19, 20]. For example, the modification $\Delta b_X(z,k)$ to the Gaussian large-scale bias $b_X^G$ of a biased tracer $X$ induced by local non-Gaussianity reads $\Delta b_X(z,k) = 3[b_X^G(z) - 1]\Omega_m {H_0}^2\delta_c/[k^2T(k)D_+(z)]f_{\rm NL}$, where $\Omega_m$ is the total mass fraction in units of the critical density, $H_0$ the Hubble constant, $\delta_c$ the critical value of the matter over-density at collapse, $T(k)$ the transfer function, $D_+(z)$ the linear growth factor of density perturbations normalised to unity today and $f_{\rm NL}$ parameterises the quadratic correction to the linear Gaussian term in Bardeen's gauge invariant potential [21, 22]. Attempts at detecting this effect with galaxy surveys already led to promising results (e.g. [23]). In what follows, we shall exploit this peculiar behaviour to forecast constraints on primordial non-Gaussianity with HI intensity mapping.

## 3. HI intensity mapping

Late-time cosmology with HI intensity mapping is based on the assumption that, after reionisation, most of HI in the cosmos is confined within galaxies. Thus, a map of its 21 cm line emission is a proxy of the underlying clustered distribution of galactic haloes. The idea is that radio telescopes measure flux density. In the Rayleigh-Jeans limit, this can be converted into an effective brightness temperature – in turn split into a homogeneous and a fluctuating part, $T_b = \overline{T}_b(1+\delta_{\rm HI})$, where $\delta_{\rm HI}$ are the perturbations in the large-scale HI distribution. By investigating the statistical properties of such HI over-densities in the Universe, we can thus reconstruct the fluctuations of underlying matter density field, much in the same way as it is done with the power spectrum or the correlation function of usual galaxy surveys.

There are two other main ingredients for constructing the power spectrum of HI density fluctuations, namely its bias, $b_{\rm HI}$, and the total HI density fraction, $\Omega_{\rm HI}$. The former depends upon the size of host dark matter haloes. Such dependence can be modelled by using the halo mass function with an appropriate lower mass cutoff (see e.g. [24–26] for more discussions). Despite the considerable disagreement between various HI bias models, we have got at least one robust measurement of the combined quantity $\Omega_{\rm HI}b_{\rm HI} = (4.3 \pm 1.1) \times 10^{-4}$ at the 68% confidence level at $z = 0.8$ [25]. This constraint can be used to gauge a semi-analytical derivation of $b_{\rm HI}$ and $\Omega_{\rm HI}$ [27, 28]. Firstly, we need to specify the amount of HI within a halo of mass $M$ at redshift $z$, $M_{\rm HI}(M, z)$. The most straightforward ansatz is a direct proportionality with respect to the halo mass, which can be then fitted against the available data. Thence, we can calculate $\Omega_{\rm HI}$, the HI bias and HI brightness temperature in a consistent manner.

The principal advantage of HI intensity mapping experiments in the task of surveying the cosmic structure on the largest scales is that they are sensitive to structures at a redshift range which is difficult to span for conventional galaxy surveys [29]. Indeed, for them it is hard to achieve the required sensitivity at high redshift over vast areas of the sky. Moreover, the problem of a poor sampling, usually referred to as cosmic variance, can greatly spoil the results. However, by not requiring galaxy detections, the intensity mapping technique transfers the problem to the issues of calibration and foreground removal, viz. the need for cleaning methods for removing everything but the HI signal we are interested in [30].

## 4. Forecasts for future surveys

Now, we show forecast constraints on the primordial non-Gaussianity parameter, $f_{\rm NL}$, as achievable by viable future experiments. If we work with Fourier-Bessel transforms on the celestial sphere, the HI angular power spectrum is $C_\ell^{\rm HI}(\nu_i, \nu_j)$, where $\nu_i$ is the frequency of $i$th shell. To calculate this quantity, we use the CAMB_sources code [9], and include the redshift space distortion but discard subdominant terms. We refer to Camera et al. [1] for other assumptions and the experimental set up. Fig. 1 from Ref. [1] depicts the main results. We plot $\sigma_{f_{\rm NL}}$ contours in the plane of surveyed area and total observation time, with abscissas covering from a $15 \times 15\,{\rm deg}^2$ survey to $2\pi$. The three top panels, where the $y$-axis is observation time $t_{\rm TOT}$ multiplied by the number of dishes $N_d$, illustrates the case for a dish survey with three distinct maximum angular modes, corresponding to dish diameters of 5, 15 and 80 metres at redshift $\sim 3$. For higher angular resolution, interferometers may be a better option. The lower panels of Fig. 1 show $\sigma_{f_{\rm NL}}$ for 1, 10 and 100 pointings at a resolution of $\ell_{\rm max} \simeq 300$, as set by choosing $D_a \sim 80\,{\rm m}$ as the diameter for the array. For interferometers, the main design parameter is the field of view, which fixes $\ell_{\rm min}$, and for a 'dense array' it is related to the number of elements. Given that the maximum angular scale is set by the field of view, by pointing several times we simply diminish the variance by $N_p$, though the noise increases too, because $t_{\rm obs} \to t_{\rm obs}/N_p$.

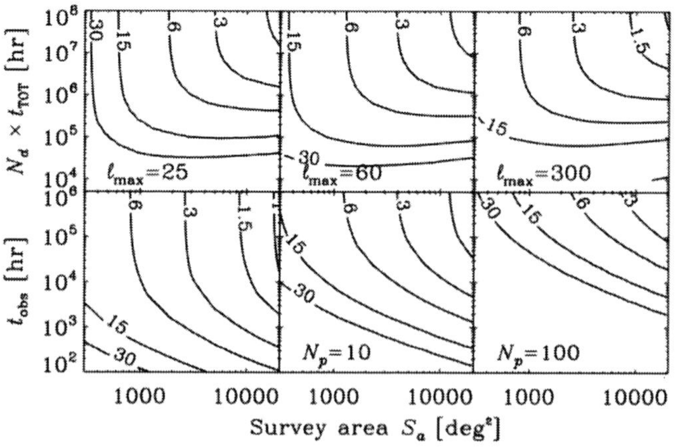


**Figure 1.** Forecast 68.3% error contours on $f_{\rm NL}$ as a function of surveyed area and total observation time, for a dish survey with $N_d$ dishes (upper panels) and an interferometer making $N_p$ pointings (lower panels). Courtesy of [1].

Several dish survey telescopes under development should be able to probe this HI intensity signal and thus constrain primordial non-Gaussianity. One of the major candidate is the Square Kilometre Array (SKA) phase 1, which will probe down to $\sim 350\,{\rm MHz}$, thus allowing us to push below *Planck* constraints. In terms of designing a new system from scratch, something like 10,000 small dishes between $2 - 4\,{\rm m}$

diameter, working at $\sim$ 400 MHz, would be a good and cheap possibility to target the $f_{\rm NL} \sim 1$ region. For interferometric surveys, none of the planned telescopes are compact enough to deliver the required sensitivity on large scales. This means we would need in principle to wait for the full SKA, with the proposed 'aperture array' system working below 1 GHz, thus reaching the $\sigma_{f_{\rm NL}} \lesssim 1$ limit. However, the full SKA is designed to achieve much higher angular resolution than what is needed for our purposes and a smaller array with 80 m or less in diameter would be an interesting, near term, alternative, capable of reaching $\sigma_{f_{\rm NL}} \sim 1$.

## 5. Conclusions

Here, we have succinctly reviewed the current theoretical and experimental endeavour to the aim of accessing the largest cosmological scales. We have shown why such extremely large scales contain a wealth of information capable to deepen our understanding of the physical mechanism driving the early Universe inflation and of the law of gravity itself. We have presented the HI intensity mapping technique as a valuable tool to access ultra-large scales – hardly reachable with enough sensitivity for conventional galaxy surveys. As a figure of merit, we have focussed on primordial non-Gaussianity as an emblematic large-scale effect and have shown how forthcoming HI intensity mapping experiments will enable us to constrain $f_{\rm NL}$ to a degree of accuracy much higher than the current most stringent constraints achieved by the Planck CMB experiment.

**Acknowledgments**
SC wishes to thank his collaborators for allowing him to show results from common projects. SC is funded by FCT-Portugal under Post-Doctoral Grant No. SFRH/BPD/80274/2011.

## References

[1] Camera S, Santos M G, Ferreira P G and Ferramacho L 2013 *Phys. Rev. Lett.* **111** 171302 (*Preprint* 1305.6928)
[2] Camera S *et al.* 2014 *Advancing Astrophysics with the Square Kilometre Array*
[3] Santos M G *et al.* 2014 *Advancing Astrophysics with the Square Kilometre Array*
[4] Bertschinger E 1998 *Ann.Rev.Astron.Astrophys.* **36** 599–654
[5] Dolag K, Borgani S, Schindler S, Diaferio A and Bykov A M 2008 *Space Science Reviews* **134** 229–268
[6] Van Daalen M P, Schaye J, Booth C M and Dalla Vecchia C 2011 *Mon. Not. Roy. Astron. Soc.* **415** 3649–3665
[7] Yoo J 2010 *Phys. Rev.* **D82** 083508 (*Preprint* 1009.3021)
[8] Bonvin C and Durrer R 2011 *Phys. Rev.* **D84** 063505 (*Preprint* 1105.5280)
[9] Challinor A and Lewis A 2011 *Phys. Rev.* **D84** 043516 (*Preprint* 1105.5292)
[10] Clifton T, Ferreira P G, Padilla A and Skordis C 2012 *Phys. Rept.* **513** 1–189 (*Preprint* 1106.2476)
[11] Bartolo N, Komatsu E, Matarrese S and Riotto A 2004 *Phys. Rept.* **402** 103–266 (*Preprint* astro-ph/0406398)
[12] Sasaki M, Valiviita J and Wands D 2006 *Phys. Rev.* **D74** 103003 (*Preprint* astro-ph/0607627)
[13] Bartolo N, Matarrese S and Riotto A 2002 *Phys. Rev.* **D65** 103505 (*Preprint* hep-ph/0112261)
[14] Arkani-Hamed N, Creminelli P, Mukohyama S and Zaldarriaga M 2004 *JCAP* **0404** 001 (*Preprint* hep-th/0312100)
[15] Creminelli P, Senatore L, Zaldarriaga M and Tegmark M 2007 *JCAP* **0703** 005 (*Preprint* astro-ph/0610600)
[16] Chen X, Huang M x, Kachru S and Shiu G 2007 *JCAP* **0701** 002 (*Preprint* hep-th/0605045)
[17] Senatore L, Smith K M and Zaldarriaga M 2010 *JCAP* **1001** 028 (*Preprint* 0905.3746)
[18] Ade P *et al.* (Planck Collaboration) 2013 (*Preprint* 1303.5084)
[19] Dalal N, Dore O, Huterer D and Shirokov A 2008 *Phys. Rev.* **D77** 123514 (*Preprint* 0710.4560)
[20] Matarrese S and Verde L 2008 *Astrophys. J.* **677** L77–L80 (*Preprint* 0801.4826)
[21] Verde L, Wang L M, Heavens A and Kamionkowski M 2000 *Mon. Not. Roy. Astron. Soc.* **313** L141–L147
[22] Komatsu E and Spergel D N 2001 *Phys. Rev.* **D63** 063002 (*Preprint* astro-ph/0005036)
[23] Giannantonio T, Ross A J, Percival W J, Crittenden R, Bacher D *et al.* 2014 *Phys. Rev.* **D89** 023511 (*Preprint* 1303.1349)
[24] Bagla J, Khandai N and Datta K K 2010 *Mon. Not. Roy. Astron. Soc.* **407** 567 (*Preprint* 0908.3796)
[25] Switzer E R, Masui K W, Bandura K *et al.* 2013 *Mon. Not. Roy. Astron. Soc.* **434** L46–L50
[26] Jarvis M J, Bhatnagar S, Bruggen M, Ferrari C, Heywood I *et al.* 2014 (*Preprint* 1401.4018)
[27] Gong Y, Chen X, Silva M, Cooray A and Santos M G 2011 *Astrophys. J.* **740** L20 (*Preprint* 1108.0947)
[28] Bull P, Ferreira P G, Patel P and Santos M G 2014 (*Preprint* 1405.1452)
[29] Seo H J, Dodelson S, Marriner J, Mcginnis D, Stebbins A *et al.* 2010 *Astrophys.J.* **721** 164–173 (*Preprint* 0910.5007)
[30] Alonso D, Ferreira P G and Santos M G 2014 (*Preprint* 1405.1751)

Journal of Physics: Conference Series **566** (2014) 012005 doi:10.1088/1742-6596/566/1/012005

# Dark Energy coupling with electromagnetism as seen from future low-medium redshift probes

**Stefania Pandolfi**

Dark Cosmology Centre, Niels Bohr Institute, University of Copenhagen, Julianes Maries Vej, 2100, Copenhagen, Denmark

E-mail: stefania@dark-cosmology.dk

**Abstract.** Beyond the standard cosmological model where the late-time accelerated expansion of the universe is driven by a cosmological constant, the observed expansion can be reproduced as well by the introduction of an additional dynamical scalar field. In this case, the field is expected to be naturally coupled to the rest of the theory's fields, unless a (still unknown) symmetry suppresses this coupling. Therefore, this would possibly lead to some observational consequences, such as space-time variations of nature's fundamental constants. In this paper we investigate the coupling between a dynamical Dark Energy model and the electromagnetic field, and the corresponding evolution of the fine structure constant ($\alpha$) with respect to the standard local value $\alpha_0$.

## 1. Introduction

Since the discovery of cosmic acceleration from measurements of luminosity distances of type Ia Supernovae (SN) in 1998 [1, 2] and its confirmation by several other independent cosmological data, the nature of the component driving this acceleration, the so-called Dark Energy (DE hereafter), has been deeply debated. In the standard cosmological model, the $\Lambda$ Cold Dark Matter ($\Lambda$CDM), the acceleration is produced by the cosmological constant $\Lambda$. This model is consistent with the majority of the observational data, but the known theoretical problems of the cosmological constant led cosmologists to formulate several other alternative models able, from one side, to relieve the already mentioned theoretical issues and, on the other side, to explain observations.

Alternative models for the DE, such as quintessence, are called dynamical dark energy and, even if not favored, they are currently not excluded by observations [3]. Several of these alternative models are characterized by the existence of an additional scalar field which drives the accelerated expansion of the universe. If this is the case, it is expected that this additional component is coupled to the rest of the theory's fields. In other words, the dynamical scalar fields are expected to be naturally coupled to the rest of the theory, unless a (still unknown) symmetry suppresses this coupling [4].

In this paper we study the coupling of dynamical DE models with the electromagnetic field: the presence of this coupling would lead to a space-time variation of the fine-structure constant $\alpha$ [4]. This, in turn, would generate distinctive signatures in cosmological data, such as the Cosmic Microwave Background (CMB), but also in low and medium redshift cosmological probes, for example in the peak of luminosity in SN or in the metal absorption lines of distant quasars

(QSO). The relevance of this combination of probes is the coverage of a wide redshift range ($0 < z \leq 5$) which is a very powerful way to discriminate between a cosmological constant and a dynamical DE model, as it makes possible to investigate the onset of DE.

## 2. Theoretical Models

We assume the Dark Energy to be a quintessence component, coupled to the electromagnetic sector, thus leading to the time variation of the fine structure constant $\alpha$. We consider a phenomenological generic parametrization of the DE equation of state parameter: the Chevallier-Polarski-Linder (CPL) parametrization [5, 6].

In the CPL model the DE equation of state (EoS) is written as $w_{\rm CPL}(z) = w_0 + w_a(z/(1+z))$ where $w_0$ is the present value of $w_{\rm CPL}$ (i.e. $w_{\rm CPL}(z=0) = w_0$) and $w_a$ is the coefficient of the time-dependent term of the EoS. In this model the EoS has a trend with redshift that is not intended to reproduce a particular model for dark energy, but rather to allow to probe possible deviations from the ΛCDM standard paradigm without the assumption of any underlying theory. Nevertheless, we can assume that also this kind of DE is produced by a scalar field. Our aim is to study the coupling of the dark energy degree of freedom with the electromagnetic field. It can be shown that the coupling between the scalar field, $\phi$, and electromagnetism possibly leads to the evolution of $\alpha$ which is given by :

$$\frac{\Delta\alpha}{\alpha}(z) = \zeta \int_0^z \sqrt{3\Omega_\phi(z)\,[1+w(z)]}\frac{dz'}{1+z'}\,. \tag{1}$$

As expected, in this class of models the magnitude of the $\alpha$ variation is controlled by the strength of the coupling $\zeta$. Here $\Omega_\phi(z)$ is the fraction of energy density provided by the scalar field, and $\Omega^0_{\rm m}$ and $\Omega^0_{\rm CPL}$ are, respectively, the present time energy densities of matter and DE.

## 3. Observational Probes

### *3.1. Supernovae Type Ia data*

Type Ia Supernovae are bright, standardizable candles, and can be used to constrain cosmic acceleration through the Hubble diagram. At present, they are the most effective and mature probe of dark energy. Moreover, as the SN peak luminosity ($L_{peak}$) depends on photon diffusion time, which in turn depends on $\alpha$ through the opacity, the $\alpha$ variation could affect $L_{peak}$ [8]. Decreasing alpha decreases the opacity, allowing photons to escape faster, thus increasing $L_{peak}$. This can be translated (see[8]) to a change in the distance modulus $\mu = m - M$, with $m$ the apparent magnitude, as

$$\mu(z) = m - M = m - (M_0 + \Delta M) = \mu_0(z) - \frac{1}{0.98}\frac{\Delta\alpha}{\alpha}(z) \tag{2}$$

where $\mu_0(z) = 5\log_{10}(d_L(z)) + 25$ is function of the luminosity distance, whose expression encodes the chosen dark energy model. We build the SN datasets following the procedure presented in [9], using Euclid specifications [10, 11] to forecast a SN survey at low-intermediate $z$, containing 1700 supernovae uniformly distributed in the redshift range $0.75 < z < 1.5$.

### *3.2. Quasar absorption systems data*

The frequencies of narrow metal absorption lines in quasar absorption systems are sensitive to $\alpha$ [12], and the different transitions have different sensitivities. Observationally, one expects relative velocity shifts between transitions in a given absorber, in a single spectrum, if $\alpha$ does vary; this comparison can therefore be used to obtain measurements of $\alpha$ in these absorption systems. Indeed a survey able to observe quasar absorption lines at different redshifts is able to reconstruct the variation of $\alpha$ with respect to the present value and to provide a dataset corresponding to the left-hand side of Eq. (1). For representative future datasets we use the baseline (conservative) case discussed in [13]. We consider the European Extremely Large Telescope (E-ELT) equipped with a high-resolution, ultra-stable spectrograph (ELT-HIRES), for which the COsmic Dynamics Experiment (CODEX) Phase A study [14] provides a baseline reference. We assume uniformly distributed measurements in the redshift range $0.5 < z < 4.0$, with an error $\sigma_\alpha = 10^{-7}$.

### *3.3. Redshift-drift data*

QSO observations can be also used to constrain DE models through the so called redshift-drift of these sources [15, 16]. The redshift-drift is the change of the redshift due to the expansion of the universe between two observations of the same distant source spectrum, repeated after a given amount of (terrestrial) years. The required time lapse depends on the instrument used (and specifically on its calibration stability) but is typically of the order of a decade with next-generation facilities.

With this kind of observations one can exploit distant astrophysical sources as a probe of the expansion of the universe in a model independent way [17, 18, 19]. As pointed out in [20, 7] QSO are the ideal astrophysical objects to observe the redshift variation $\Delta z$ between two observations. This $\Delta z$ can be translated to a spectroscopic velocity $\Delta v = c\Delta z/(1+z)$ and connected to cosmological quantities through the relation

$$\frac{\Delta v}{c} = H_0 \Delta t \left[1 - \frac{E(z)}{1+z}\right], \tag{3}$$

where $c$ is the speed of light, $\Delta t$ is the time interval between two observations of the same astrophysical source, and $E(z) = H(z)/H_0$ where $H(z)$ is the Hubble parameter and $H_0$ is it's value today, is expression encodes the chosen dark energy model.

A CODEX-like spectrograph on the E-ELT will have the ability to detect the cosmological redshift-drift in the Lyman $\alpha$ absorption lines of distant ($2 < z < 5$) QSOs, in a period of $\Delta t = 20$. According to [14], the error on the measured spectroscopic velocity shift $\Delta v$ that can be expressed as:

$$\sigma_{\Delta v} = 1.35 \, \frac{2370}{S/N} \sqrt{\frac{30}{N_{\rm QSO}}} \left(\frac{5}{1+z_{\rm QSO}}\right)^x \; cm \; s^{-1}, \tag{4}$$

where $S/N$ is the signal to noise ratio, $N_{\rm QSO}$ the number of observed quasars, $z_{\rm QSO}$ their redshift and the exponent $x$ is equal to 1.7 when $z \leq 4$, while it becomes 0.9 beyond that redshift.

Therefore, we can forecast a redshift-drift dataset where the error bars are computed using Eq.(4), with $S/N = 3000$ and a number of QSO $N_{\rm QSO} = 30$ is assumed to be uniformly distributed among the following redshift bins $z_{\rm QSO} = [2.0, 2.8, 3.5, 4.2, 5.0]$.

### *3.4. Weak lensing data*

Weak gravitational lensing of distant galaxies is a powerful observable to probe the geometry of the universe and to map the dark matter distribution. We simulate a weak lensing dataset according to the specifications expected for the Euclid survey [10]: the mission will observe $n_g \simeq 30$ gal/arcmin$^2$ over an area $\Omega = 15000$ deg$^2$, corresponding to a sky fraction $f_{sky} \sim 33\%$. The large galaxy number density and the wide area observed will allow Euclid to provide us with a tomographic reconstruction of the weak lensing signal. We therefore divide the redshift space in 10 bins, chosen in such a way to have the same fraction of the total observed galaxies in each one. Using these specifications we build the $\ell$-by-$\ell$ convergence power spectrum and the $1\sigma$ uncertainties, computed as in [21, 22].

## 4. Analysis

We build simulated datasets assuming a fiducial cosmology given by the observations of the WMAP satellite after 9 years of data [23] for the standard parameters. We fix the DE parameters in such a way to mimic the ΛCDM expansion (i.e. $w_0 = -1$, $w_a = 0$) and a vanishing coupling $\zeta = 0$.

In a second case, we also build simulated datasets with a non vanishing variation of $\alpha$ assuming the same value of the previous case for the standard parameters, but different values for the ones involved in the $\alpha$ variation.

We rely on a MCMC analysis to sample the parameter space and we use a modified version of the publicly available MCMC package `cosmomc` [24] with a convergence diagnostic using the Gelman and Rubin statistics. We assume flat priors on the sampled parameters.

## 5. Results

As stated in the previous section, the first investigation we carry out deals with vanishing $\Delta\alpha/\alpha$ mock datasets. We consider different combinations of the probes introduced in Section 3 and discuss the main features obtained by this analysis, exploring how the main geometrical probes (WL and SN) affect constraints on DE parameters and on the coupling $\zeta$.

In Fig. 1 we can notice how the Euclid survey will greatly narrow the allowed parameter space for the EoS parameters $w_0$ and $w_a$, mainly thanks to the combination of the SN and WL measurements. When we consider all datasets we get $\sigma(w_0) = 0.007$ and $\sigma(w_a) = 0.03$.

The constraints on the coupling parameter are instead puzzling at a first look (see last panel in Fig. 1), as the use of the Euclid observations loosens the bounds on $\zeta$. This result is however

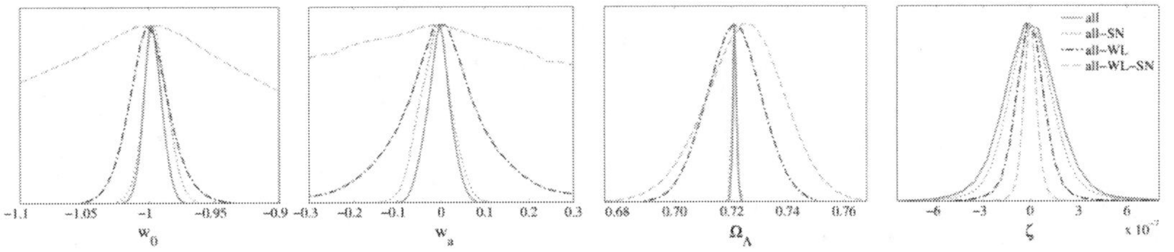


**Figure 1.** Marginalized 1-dimensional posterior distributions for the DE parameters $w_0$, $w_a$, $\Omega_\Lambda$ and the coupling $\zeta$, for different combinations of probes.

easily explained considering the chosen fiducial cosmological model. Eq.(1) in fact implies that a vanishing $\Delta\alpha/\alpha$ can be obtained in two ways: either $\zeta = 0$ and/or $w(z) = -1$. This leads to the fact that when $w_0$ and $w_a$ are poorly constrained (i.e. when WL and SN are removed from the analysis) the QSO forecasted measurements require a coupling $\zeta$ close to zero. On the contrary when WL and SN impose tight independent constraints on DE parameters and the recovered $w(z)$ is close to $-1$, a larger range of $\zeta$ values is in agreement with the QSO measurements.

In a second step of our analysis we select a fiducial model where $\Delta\alpha/\alpha$ is not vanishing and the DE parameters move from the standard ΛCDM scenario. In this case, the peculiar $w - \zeta$ behavior mentioned above, due to the $\zeta = 0$ fiducial value, is not present and the degeneracies between these parameters show up clearly, as we report in Fig. 2.

We also notice that probing a different fiducial cosmology will give different constraints on the parameters. The constraint on $w_0$ improves by a factor of about two and the measurement of $w_a$ becomes about one order of magnitude better: moving the fiducial region away from the special point ($\zeta = 0$, $w_0 = -1$) prevents the loss of constraining power because of the

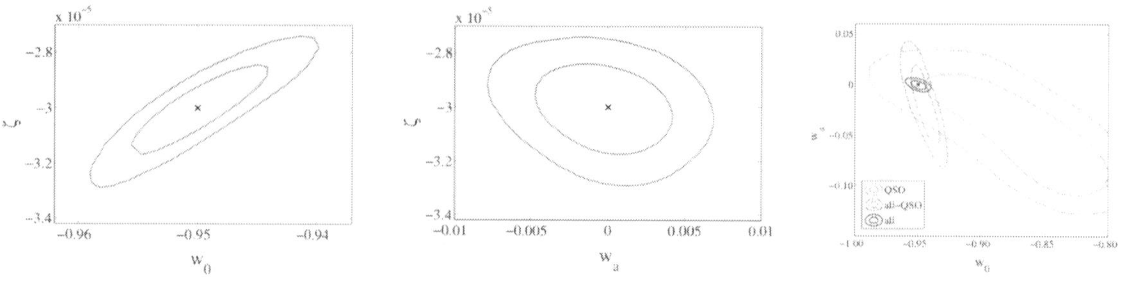


**Figure 2.** *Left and middle panel* : 2-dimensional contours at 68% and 95% confidence levels showing $\zeta$ versus $w_0/w_a$ for the CPL model. *Right panel* QSO contribution to the $w_0$-$w_a$ constraints. We report contour plots at 68% and 95 % confidence levels as obtained from QSO data only (dashed green line), all probes except QSO (dash-dotted red line) and all probes (solid purple line). The black cross shows the fiducial input values.

pathological degeneracies described in Fig. 1 and therefore all the observables can fully contribute in constraining the cosmological parameters. In particular, in these non standard scenarios, the QSO contribution will be non vanishing. Even though QSO data have a much lower constraining power than other dark energy observables, in the rightmost panel of Fig. 2, it is possible to notice how this dataset can provide independent (and almost orthogonal) limits on dark energy parameters and can be used to break degeneracies between $w_0$ and $w_a$.

## References

[1] Perlmutter S *et al.* [Supernova Cosmology Project Collaboration] 1999 *Astrophys. J.* **517** 565.
[2] Riess A G *et al.* [Supernova Search Team Collaboration] 1998, *Astron. J.* **116** 1009.
[3] Ade P A R *et al.* [ Planck Collaboration] 2013, arXiv:1303.5076 [astro-ph.CO].
[4] Carroll S M 1998 *Phys. Rev. Lett.* **81** 3067.
[5] Chevallier M, Polarski D, 2001 *Int. J. Mod. Phys.* D **10** 213 [gr-qc/0009008].
[6] E. V. Linder 2003 *Phys. Rev. Lett.* **90** 091301
[7] Martinelli M, Pandolfi S, Martins C J A P and Vielzeuf P E 2012, *Phys. Rev.* D **86** 123001
[8] Chiba T and Kohri K 2003 *Prog. Theor. Phys.* **110** 195
[9] Cardone V F , Spiro S, Hook I and Scaramella R 2012, *Phys. Rev.* D **85** 123510
[10] Laureijs R, Amiaux J, Arduini S, Augueres J L, Brinchmann J, Cole R, Cropper M and Dabin C *et al.*, arXiv:1110.3193 [astro-ph.CO].
[11] Hook I M, 2012 arXiv:1211.6586 [astro-ph.CO].
[12] Bahcall J N, Sargent W L and Schmidt M, 1967 *Astrophys. J.* **149** L11
[13] Amendola L, et al. , 2012 *Phys. Rev.* D **86** 063515
[14] CODEX Phase A Science Case, 2010 document E-TRE-IOA-573-0001 Issue 1
[15] Sandage A, 1962 *Ap.J.* **136**, 319-33
[16] Loeb A, 1998 *Astrophys. J.* **499** L111
[17] Pasquini L , et al., 2005, *The Messenger*, 122, 10
[18] Corasaniti P S, Huterer D and Melchiorri A, 2007 *Phys. Rev.* D **75** 062001
[19] Quercellini C, Amendola L, Balbi A, Cabella P and Quartin M, 2012 *Phys. Rept.* **521** 95
[20] Vielzeuf P E and Martins C J A P, 2012 *Phys. Rev.* D **85** 087301
[21] De Bernardis F, Martinelli M, Melchiorri A, Mena O and Cooray A, 2011 *Phys. Rev.* D **84**, 023504
[22] Cooray A R, 1999 *A&A* , **348**, 31 (1999)
[23] Hinshaw G *et al.* [WMAP Collaboration], 2012
[24] Lewis A and Bridle S, 2002 *Phys. Rev.* D **66** 103511
[25] Liske J, Grazian A, Vanzella E, Dessauges M, Viel M, Pasquini L, Haehnelt M and Cristiani S *et al.*, 2008 *Mon. Not. Roy. Astron. Soc.* **386**, 1192

5th Young Researcher Meeting, Trieste 2014 IOP Publishing
Journal of Physics: Conference Series **566** (2014) 012006 doi:10.1088/1742-6596/566/1/012006

# Time-evolution of the fine-structure constant in runaway dilaton models

**Pauline Eva Vielzeuf and C. J. A. P. Martins**

Centro de Astrof sica da Universidade do Porto, Rua das Estrelas, 4150-762 Porto, Faculdade de Ciencias, Universidade do Porto, Rua do Campo Alegre 687, 4169-007 Porto, Portugal

E-mail: pauline.vielzeuf@astro.up.pt, carlos.martins@astro.up.pt

**Abstract.** We study the detailed time-evolution of the fine-structure constant $\alpha$ in the string-inspired runaway dilaton class of models of Damour, Piazza and Veneziano [1, 2]. We provide constraints on this scenario via the time-variations of the fine-structure constant $\alpha$ as measured by spectroscopic experiments and we explore ways to distinguish the dilaton runaway models from other alternative.

## 1. Introduction

One of the main open questions in cosmology is the acceleration of the expansion of the Universe. The simplest model, a constant energy term, the so-called cosmological constant, fits the data very well, but it is unsatisfactory from a theoretical point of view.

An alternative way to model cosmic acceleration as coming from a dynamical energy component is through a scalar field, similar in the mathematical description to the Higgs boson, recently discovered at the Large Hadronic Collider. Such scalar fields emerge quite naturally in string theory, which predicts the presence of a scalar partner of the spin-2 graviton, the dilaton. We will focus our analysis on the cosmological consequences of a particular class of string-inspired models, the runaway dilaton scenario of Damour, Piazza and Veneziano [1, 2] and assess their testability by future facilities. Specifically ELT-HIRES [3] an ultra-stable spectrograph for the E-ELT (European Extremely Large Telescope) will have two relevant capabilities: a direct measurement of the cosmic expansion performing then the so-called Sandage-Loeb test [3, 4], and tests of the stability of the fine-structure constant ($\alpha = \frac{e^2}{\hbar c}$ ) at up to $10^{-8}$ level.

## 2. Runaway Dilaton Cosmology

The Friedmann equation and evolution equation for the scalar field ($\Phi$) for this class of models can be written as:

$$3H^2 = 8\pi G \sum_i \rho_i + H^2 \Phi'^2 \tag{1}$$

$$\frac{2}{3-\Phi'^2}\Phi'' + \left(1 - \frac{p}{\rho}\right)\Phi' = -\sum_i \alpha_i(\Phi)\frac{\rho_i - 3p_i}{\rho}. \tag{2}$$

The total pressure $p = \sum_i p_i$ and the total energy density $\rho = \sum_i \rho_i$ sum over all components except the kinetic part of the scalar field, $\alpha_i(\Phi)$ are the coupling constants between the dilaton and each component $i$, so they characterize the effect of the various components of the universe on the dynamics of the field. Notice in particular that the theory does not require the coupling constant to be the same for all components. Experimental constraints impose a tiny coupling to baryonic matter:

$$\alpha_{had} \sim 40 b_F c e^{-c\Phi} \tag{3}$$

where $b_F$ and c are constant free parameters (c is expected to be of order unity and $b_F$ very small). Weak equivalence principle tests lead to the following bound on the present value of the coupling:

$$|\alpha_{had,0}| \leq 10^{-4} \tag{4}$$

leading to the following relation between $b_F$ and $\Phi_0$:

$$\Phi_0 \geq \ln\frac{|b_F|}{2 \times 10^{-}6}. \tag{5}$$

Using the basic definition of deceleration parameter $q = -1 - \frac{\dot{H}}{H^2}$ one can also derive the constraint:

$$|\Phi_0'| \leq 0.3 \tag{6}$$

where the dots represent the derivative with respect to time, and the prime with respect to the logarithm of the scale factor ln(a).

By solving the Friedmann equations for these classes of models one finds the redshift evolution of the dilaton field compared to the present value of the field (fig.1) and the evolution of the Hubble parameter for these type of model (fig.2) which should be compared to observational data. A comprehensive list of recent measurements is given in [7].

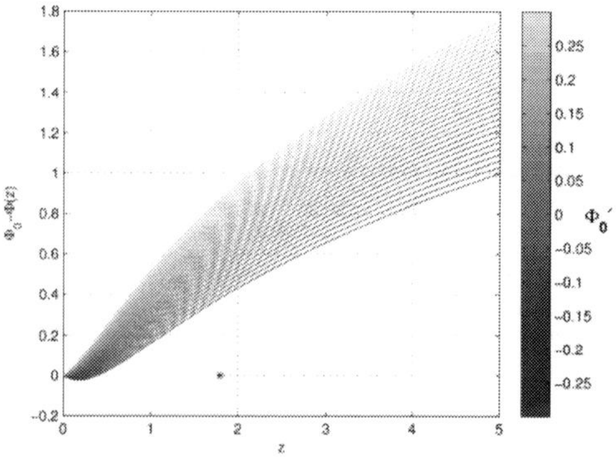


**Figure 1.** Redshift evolution of dilaton field compare to its present value.

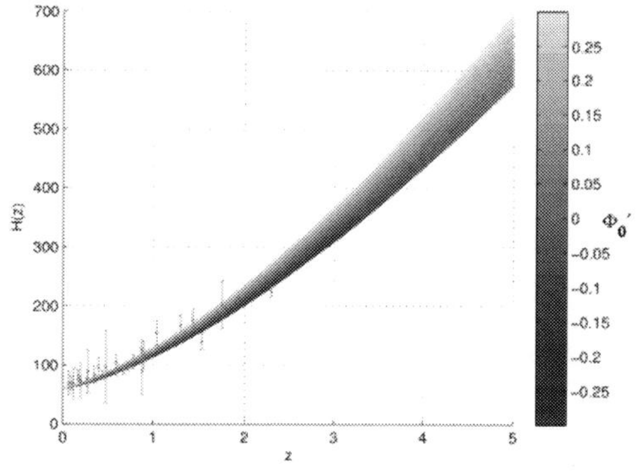


**Figure 2.** Hubble parameter as a function of redshift of the dilaton models with experimental data from [7] in both panel, $b_f$ had been fixed to $10^{-8}$

## 3. Variations of fundamental constants

Recent obsevations suggest possible variations of fine-structure constant in time and/or space [8]. Our analysis is focussed on time-variations of $\alpha$ and some specific measurements are in Table 1, including the recent first result of the UVES Large Program for Testing Fundamental Physics [5], which is expected to be the one with a better control of possible systematics.

Journal of Physics: Conference Series **566** (2014) 012006 doi:10.1088/1742-6596/566/1/012006

**Table 1.** Available specific measurements of $\alpha$, with one-sigma uncertainties.

| Object | Redshift | $\Delta\alpha/\alpha$ | Spectrograph |
|---|---|---|---|
| HE0515-4414 | 1.15 | $(-0.1 \pm 1.8) \times 10^{-6}$ | UVES |
| HE0515-4414 | 1.15 | $(0.5 \pm 2.4) \times 10^{-6}$ | HARPS+UVES |
| HE0001-2340 | 1.58 | $(-1.5 \pm 2.6) \times 10^{-6}$ | UVES |
| HE2217-2818 | 1.69 | $(1.3 \pm 2.6) \times 10^{-6}$ | UVES |
| Q1101-264 | 1.84 | $(5.7 \pm 2.7) \times 10^{-6}$ | UVES |

It is then interesting to study the behaviour of $\alpha$ in this class of models, and as it has been shown in [1], the evolution of $\alpha$ is given by:

$$\frac{1}{H}\frac{\dot{\alpha}}{\alpha} = \frac{b_F c e^{-c\Phi}}{1 - b_F c e^{-c\Phi}} \Phi'. \tag{7}$$

By integrating this equation, one finds the evolution of the variation of $\alpha$:

$$\frac{\Delta\alpha}{\alpha}(z) = \frac{\alpha(z) - \alpha_0}{\alpha_0} = b_F \left(e^{-c\Phi_0} - e^{-c\Phi(z)}\right). \tag{8}$$

The present drift of $\alpha$ is locally constrained by the Rosenband bound:

$$\left(\frac{1}{\alpha}\frac{d\alpha}{dt}\right)_0 = (-1.6 \pm 2.3)^{-17} yr^{-1}; \tag{9}$$

assuming the Planck value for the Hubble constant $H_0 = (67.4 \pm 1.4)\ \mathrm{kms^{-1}Mpc^{-1}}$, one can find:

$$|\alpha_{\mathrm{had},0}\Phi_0'| \sim |b_F c e^{-c\Phi_0'}| \leq 3 \times 10^{-5}. \tag{10}$$

From here one can then predict the redshift evolution of $\alpha$ taking $\Phi_0$ and $\Phi_0'$ as free parameters and make a $\chi^2$ analysis using both H(z) data from [7] and the $\alpha$ data from Table 1 (see figures (3) and (4).

## 4. Future tests

The drift in the spectroscopic velocity of an object following the Hubble flow can be obtained from the definition of redshift and expressed as:

$$\Delta v = \frac{c}{(1+z)} \Delta t (H_0(1+z) - H(z)). \tag{11}$$

Here $c$ is the speed of light and $\Delta t$ the time span of observation. The precision needed to detect this signal is expected to be reached by future facilities such as the SKA, an intensity mapping experiments, and ELT-HIRES (see [6] for the phase A study of the instrument) which will offer the unique advantage to observe this drift deep in the matter era ($z \sim 2 \rightarrow 5$) through spectroscopic measurements in the Lymann-$\alpha$ forest. ELT-HIRES is expected to reach a spectroscopic velocity precision parametrized by:

$$\sigma_v = 1.35 \left(\frac{S/N}{2370}\right)^{-1} \left(\frac{N_{QSO}}{30}\right)^{-1/2} \left(\frac{1+z_{QSO}}{5}\right)^{-1.7} \tag{12}$$

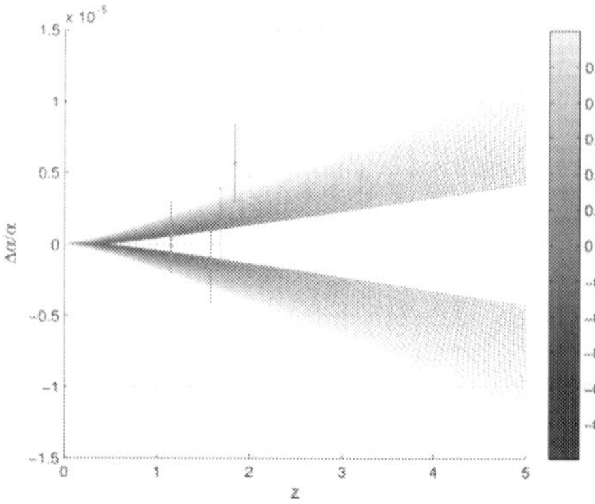


**Figure 3.** Redshift evolution of $\alpha$-variations, the data points taken from table 1.

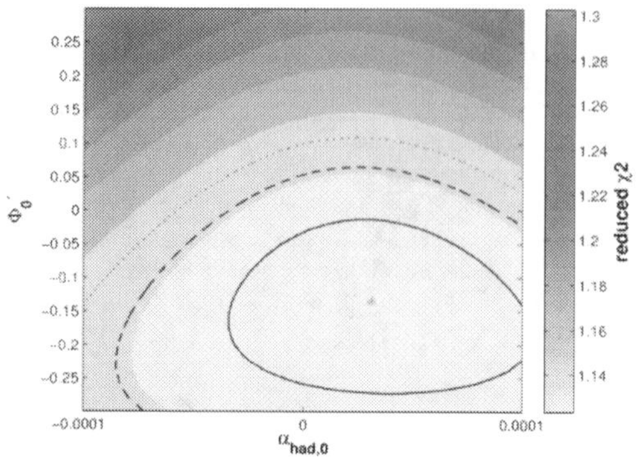


**Figure 4.** Reduced-$\chi^2$ (indicated by the color scale) using both H(z) and $\frac{\Delta\alpha}{\alpha}$ measurements, with $-10^{-8} \leq b_F \leq -10^{-9}$ (x-axis) and $-0.3 \leq \Phi_0' \leq 0.3$ (y-axis).

where $S/N$ is the signal-to-noise ratio, $N_{QSO}$ the number of targets observed, and $z_{QSO}$ the redshift of the observed targets.

Using the same values of the parameters as previously ($b_f$ fixed at $10^{-8}$, $H_0 = 67.4 \pm 1.4$ kms$^{-1}$Mpc$^{-1}$), one can then compute the behaviour of the redshift drift as a function of redshift for the runaway dilaton models, compare it to $\Lambda CDM$ models and verify whether or not one will be precise enough to put constraints on the parameter space of the model with the ELT-HIRES. Fig.5 compares the redshift drift of the standard model of cosmology (red) to the runway dilaton one, the error bars being the expected accuracy that ELT-HIRES can provide.

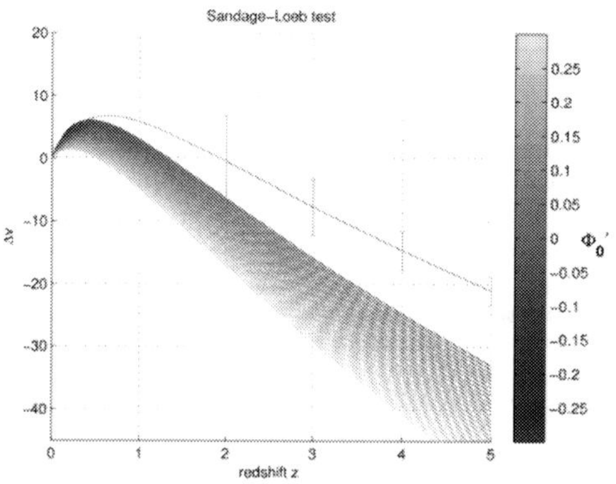


**Figure 5.** Redshift drift signal for the runaway dilaton class of models (blue), with model parameters in ranges allowed by observations, compared to the signal expected in $\Lambda CDM$ (red) and the forecasted uncertainties for an observational time of $\Delta t = 30$yrs, a signal-to-noise ratio of S/N=3000 for 40 uniformly spaced systems divided in 4 bins.

## 5. Conclusions

In this analysis, we showed that the runaway dilaton class of models remains an alternative possibility to the standard model; however, current data already places tight constraints on its free parameters. Nevertheless given these constraints, the model predicts distinctive behaviours for $\alpha$ and the redshift drift signal, which will be tested with future facilities such as ELT-HIRES.

## References

[1] T. Damour, F. Piazza and G. Veneziano. Phys. Rev. Lett., **89**:081601, 2002.
[2] T. Damour, F. Piazza and G. Veneziano. Phys. Rev. D., **66**:046007. 2002.
[3] A. Sandage. Astrophys. J. **136** (1962) 319.
[4] A. Loeb, Astrophys. J. **499** (1998), 111.
[5] P. Molaro et al., A. & A. **555**, A68 (2013) [arXiv:1305.1884 [astro-ph.CO]].
[6] J. Liske, A. Grazian, E. Vanzella, M. Dessauges, M. Viel, L. Pasquini, M. Haehnelt and S. Cristiani et al., Mon. Not. Roy. Astron. Soc. **386**, 1192 (2008) [arXiv:0802.1532[astro-ph]].
[7] O. Farooq and B. Ratra. Astrophys. J., **766**(L7), 2013.
[8] J. K. Webb, J. A. King, M. T. Murphy, V. V. Flambaum, R. F. Carswell and M. B. Bainbridge, Phys. Rev. Lett.**107**, 191101 (2011) [arXiv:1008.3907 [astro-ph.CO]].

## Acknowledgements

This work was done in the context of project PTDC/FIS/111725/2009 (FCT, Portugal). CJM is also supported by an FCT Research Professorship, contract reference IF/00064/2012, funded by FCT/MCTES (Portugal) and POPH/FSE (EC)

Journal of Physics: Conference Series **566** (2014) 012007 doi:10.1088/1742-6596/566/1/012007

# Predictions of radio counts of star forming galaxies for SKA precursors

**C Mancuso[1], A Lapi[1,2], Z-Y Cai[3], G De Zotti[1,4], F Perrotta[1], M Negrello[4] and L Danese[1]**

[1] Astrophysics Sector, SISSA, Via Bonomea 265, I-34136 Trieste, Italy
[2] Dipartimento di Fisica, Università 'Tor Vergata', Via della Ricerca Scientifica 1, I-00133 Roma, Italy
[3] Department of Astronomy and Institute of Theoretical Physics and Astrophysics, Xiamen University, Xiamen 361005, P.R. China
[4] NAF - Osservatorio Astronomico di Padova, Vicolo dell'Osservatorio 5, I-35122 Padova, Italy

E-mail: cmancuso@sissa.it

**Abstract.** We extended recent semi-analitycal galaxy formation models, that fitted the far-infrared (FIR) to millimeter wave luminosity functions and ultraviolet (UV) luminosity functions up to high redshifts, in order to take into account additional UV data and H$\alpha$ luminosity functions. The latters are tracers of star formation in evolved galaxies at low redshifts. Using the relations between star formation rate (SFR) and radio (synchrotron and free-free) emission, we obtained predictions for the counts of star-forming galaxies down to sub-nJy fluxes. These will be very useful for the new generation of radio telescopes, thanks to which it will be possible to investigate smaller and smaller radio fluxes (down to nJy), making the detection of little SFR possible.

## 1. Introduction

It is now well proven, thanks to the recent surveys carried out by *Spitzer* [1], *Herschel* [2], SCUBA[3] and South Pole Telescope (SPT) [4], that the infrared (IR) luminosity of galaxies at high-z is strongly related to ongoing star formation. However the newly born stars light is not always completely absorbed and re-emitted in the IR: this is the case of the earliest phase of galaxy evolution, when the interstellar medium is not extremely metal enriched, making the dust absorption only marginal. Also irregular and spiral galaxies at moderate to low-z, where the optical depth of the medium is little enough to let some UV radiation to survive, do not shine exclusively in the IR. The estimation of the star formation rate (SFR) is then possible thanks to a combination of IR and UV photometry, suffering from limitations in both wavebands.

Another powerful tracer of the SFR, independent of dust absorption, is the radio continuum emission. Since many years it has been established a tight relation between low frequency radio (synchrotron) and IR luminosity, even if its physical basis are not totally clear yet [5]. In fact the many physical processes contributing to this emission (propagation of relativistic electrons, strength and structure of the magnetic field, size and composition of dust grains) do not guarantee the extension of the relation up to redshift/luminosity ranges where there are not enough observational data available to test it properly.

The radio emission of star forming galaxies is however due also to a second process: the free free emission from relativistic electrons, which is directly proportional to the ionising photon rate of young, massive stars. It shows up at rest frame frequencies of tens of GHz, offering a relatively clean way to quantify the current star formation activity in galaxies. Some contamination, in this case, can derive from anomalous dust emission occurring at similar frequencies, thought to be due to spinning dust grains: there is however no proof that the contribution of this component can seriously affect the globally integrated measurements [6]. As a consequence high frequency radio observations can provide a very powerful tool to measure the star formation history of the Universe.

Padovani[7] demonstrated that at GHz frequencies, the counts below 100-200 $\mu$Jy are dominated by star-forming galaxies: unfortunately this limit is not covered by the ongoing surveys, but it will be easily reached by the new generation radio telescopes.

## 2. The model

A detailed physical model for the evolution of the UV luminosity functions, hence of the cosmic SFR at high-$z$, has been worked out by Cai[8] in the framework of the scenario proposed by Granato[9] and further elaborated by Lapi[10, 11]and Cai[12]. In the latter the authors carried out a complete investigation of the evolution of the IR luminosity functions based on a "hybrid" approach, that reflects the observed dichotomy in the ages of stellar populations of early-type galaxies on one side and late-type galaxies on the other (cf. Fig. 10 in [13]). Early-type galaxies and massive bulges of Sa type galaxies are composed of relatively old stellar populations with mass-weighted ages $\gtrsim$ 8–9 Gyr (corresponding to formation redshifts $z \gtrsim 1$–1.5), while the disc components of spirals and the irregular galaxies are characterized by significantly younger stellar populations. Thus the progenitors of early-type galaxies, referred to as proto-spheroidal galaxies or protospheroids, are the dominant star-forming population at $z \gtrsim 1.5$ while IR galaxies at $z \lesssim 1.5$ are mostly late-type "cold" (normal) and "warm" (starburst) galaxies.

The model accurately fits a broad variety of data: multi-frequency and multi-epoch luminosity functions of galaxies and Active Galactic Nuclei (AGN), redshift distributions, number counts (total and per redshift bin). It also accounts for the recently determined counts and redshift distributions of strongly lensed galaxies detected by the South Pole Telescope (SPT)[14]

Although the total (8–1000 $\mu$m) IR luminosity ($L_{IR}$) is a good proxy of the SFR when the dust heating is dominated by young stars and the effective optical/UV optical depth is high, in disks of normal galaxies the IR luminosity is the sum of a "warm" component heated by young stars and of a "cold" (or "cirrus") component, heated by the general radiation field that may be dominated by older stars. Obviously only the first component is a tracer of the SFR. This issue was investigated by Clemens[16]. We adopt the relation between SFR and $L_{\rm IR}$ derived by these authors.

Galaxies with low SFRs are also found to have low dust opacity, making the SFR derived from $L_{\rm IR}$ underestimated even by large factors. It must then be complemented using optical/UV SFR indicators: we supplemented them with a parametric phenomenological backward model, considering both the UV and the H$\alpha$ data (see Fig. 1).

For modelling the UV ($\lambda = 1500$ Å) luminosity functions of late-type galaxies we adopted the same functional form used by Cai[12] in the IR. We used a simple pure luminosity evolution model ($L^*(z) = L_0^*(1+z)^{\alpha_L}$ up to $z = 1$), that turned out to provide a sufficiently good description of the data. For proto-spheroidal galaxies we kept instead the Cai[8] physical model.

## 3. Calibration of the relation between radio emission and SFR

For extending the model to the radio frequencies we used the calibration of the relation between the SFR and the synchrotron emission calculated by Murphy[17]. In order to take into account electron ageing effects we adopted a steepening by $\Delta\alpha = 0.5$ above a break frequency of 20 GHz,

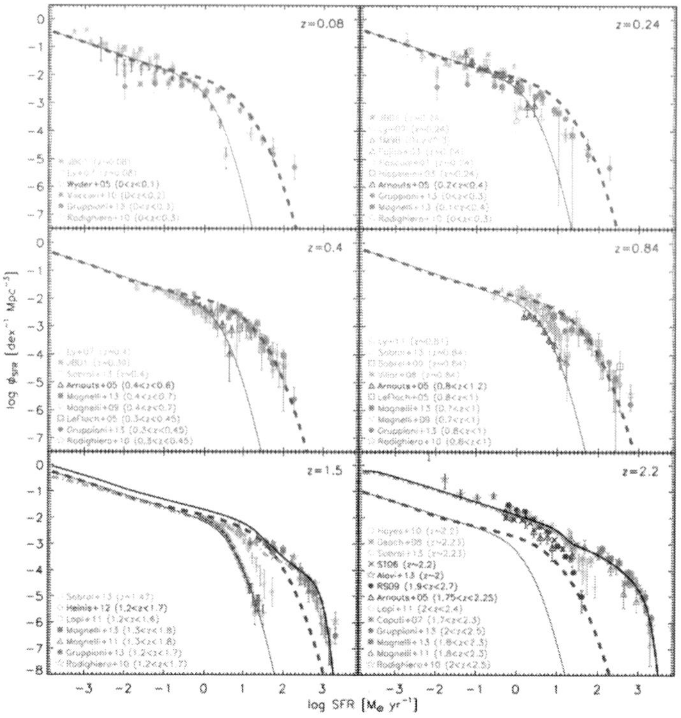


**Figure 1.** SFR functions at several redshifts. The blue lines show the SFR functions derived from the UV luminosity functions given by the phenomenological model for late-type galaxies with and without correcting the observed luminosities for extinction (dashed and thin solid lines, respectively). The dot-dashed orange lines show, only at $z \geq 1.5$, the contributions given by the physical model for proto-spheroids with a minimum halo mass $M_{\rm vir}^{\rm min} = 10^{8.5} M_{\odot}$. The thick black lines show the sum of the contributions from late-type galaxies, after the extinction correction, and of proto-spheroids.

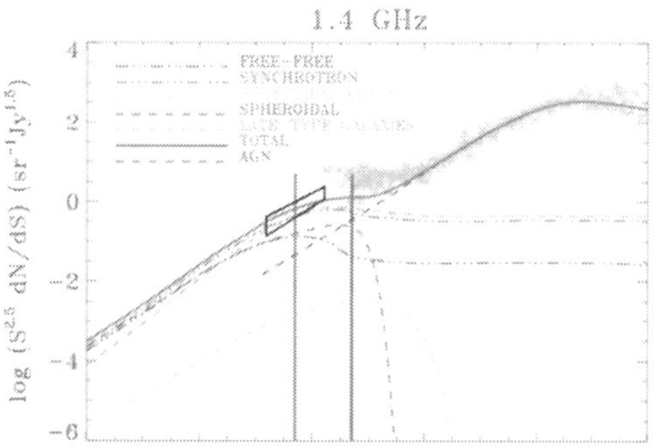


**Figure 2.** Euclidean normalized differential counts at 1.4 GHz compared with model predictions.

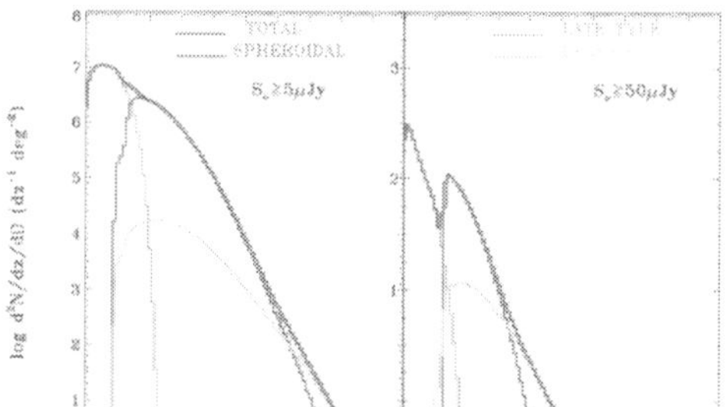


**Figure 3. Right panel:** predicted redshift distributions for 1.4 GHz surveys down to 50 and 5 $\mu$Jy.

making the SFR-synchrotron luminosity relation to be:

$$L_{\rm sync} \simeq 1.9 \times 10^{28} \left(\frac{\rm SFR}{\rm M_{\odot} yr^{-1}}\right) \left(\frac{\nu}{\rm GHz}\right)^{-0.85} \left[1 + \left(\frac{\nu}{\rm 20GHz}\right)^{0.5}\right]^{-1} \rm erg\, s^{-1}\, Hz^{-1}. \quad (1)$$

This relation was checked with the available sub-mJy 1.4 GHz observational counts (Fig. 2), showing a good match.

A relation between SFR and free-free emission was derived by Murphy[6]. We reformulated it as:

$$L_{\rm ff} = 3.75 \times 10^{26} \left(\frac{\rm SFR}{M_{\odot}/\rm yr}\right) \left(\frac{T}{10^4\,\rm K}\right)^{-0.5} \times \rm g(\nu, T) \times \exp\left(-\frac{h\nu}{k\rm T}\right) \rm erg\, s^{-1}\, Hz^{-1} \quad (2)$$

where T is the temperature of the emitting plasma and $g(\nu, T)$ is the Gaunt factor for which we adopt a more accurate approximation than that generally used [18]. The coefficient of eq. (2) was computed requiring that this equation equals that by Murphy[6] for $\nu = 33$ GHz (the frequency at which the relation was calibrated).

To test this relation we have used SPT observations of dusty galaxies at 95 GHz [14]. In fact we expect that, at this frequency, the flux densities of local galaxies not hosting a radio loud AGN are dominated by free-free emission. To estimate the 95 GHz counts of dusty galaxies Mocanu adopted a statistical approach, that is endowed with large uncertainties and may strongly overestimate the counts of dusty galaxies.We have then re-estimated the counts checking the Spectral Energy Distribution (SED) of each 95 GHz source brighter than the 95% completeness limit of 12.6 mJy, collecting all the photometric data available in the literature. We found that only 4 sources are indeed dusty galaxies, in good agreement with the modelled counts.

## 4. Predictions for new generation radio telescopes

The new generation radio telescopes include the precursors for the SKA, that are starting to run in these years. The main surveys predicted to be performed in the continuum are EMU (Evolutionary Map of the Universe), which will use the ASKAP (Australian SKA Pathfinder) telescope to make a deep ($\sim 10\mu$Jy rms) survey at 1.4 GHz, covering the entire Southern Sky, and MIGHTEE (MeerKAT International GigaHertz Tiered Exploration) on the MeerKAT telescope in South Africa, which will have $\sim 1\mu$Jy rms at 1.4 GHz [15].

We perfomed predictions for both the number counts and the redshift distributions (Figs. 2 and 3). In Fig. 2 the dashed brown line shows the best fit model for radio AGNs by Massardi[19]. The other dashed lines show the contributions of the star-forming populations considered in this paper. The two triple-dot dashed lines show the total synchrotron and free-free emissions from all the populations. We indicated with solid vertical lines the $5\sigma$ limit for both the predicted surveys (in blue the flux limit for EMU, in salmon the one for MIGHTEE). According to the model we can see that the main contributors to the "bump" at tens of $\mu$Jy levels are late-type galaxies at $z \simeq 1$–1.5. Higher $z$ proto-spheroidal galaxies become increasingly important at lower flux densities, down to a few hundred nJy's. As a result, it's clear that these new facilities will help to fill the gap between the limit reached by the current observational data and the fluxes where the contribution of star forming galaxies starts to be dominant.

The redshift distributions (Fig. 3), obtained for the $5\sigma$ flux density limits of EMU and Mightee (right and left side, respectively), show that the fraction of galaxies increases with decreasing flux density, in particular with MIGHTEE it will be possible to detect a sufficient amount of galaxies up to z=8. The tail of the distribution is moreover dominated by strongly lensed galaxies (those with magnification $\mu_{min}$=2), that could be easily distinguished from the other populations if their redshift is accurately estimated. Unfortunately it's not so easy to determine the redshifts of such objects, and we must look at some synergies with new generation telescopes, such as JWST [20], to have a reliable estimation.

## 5. Conclusions

The SKA precursors surveys can provide a much wider coverage of the SFR–$z$ plane than UV and far-IR/sub-mm surveys. We have worked out detailed predictions of the counts and redshift distributions at the fluxes that can be reached by EMU and MIGHTEE surveys, distinguishing the contributions of the different populations of star-forming galaxies. To work out such predictions we used the models by Cai[12, 8], that accurately reproduce the available determinations of the epoch-dependent SFR function from both UV and far-IR/sub-mm surveys. We combined those with the relation between SFR and radio (synchrotron and free-free) emission, obtaining an extension of the model to radio frequencies. Since the Cai models did not include the contribution to the SFR of moderate to low redshift late-type and starburst galaxies, we took into account futher information from H$\alpha$ surveys and from additional UV data in order to be as complete as possible. The relation between SFR and synchrotron emission has been checked exploiting the deepest 1.4 GHz counts, while the relation between SFR and free-free emission has been tested using the 95 GHz South Pole Telescope counts of dusty galaxies.

With this approach we have shown that the SKA precursors will start to address a flux limit region unreacheable with the current radio facilities, allowing us to have a more complete information, not affected by dust extinction, on galaxy SFRs. A more detailed discussion will be provided in a forthcoming paper [21].

## Acknowledgements

We acknowledge financial support from ASI/INAF Agreement 2014-024-R.0 for the *Planck* LFI activity of Phase E2 and from PRIN INAF 2012, project "Looking into the dust-obscured phase of galaxy formation through cosmic zoom lenses in the Herschel Astrophysical Large Area Survey".

## References

[1] Werner M. W., Roellig T. L., et al. 2004, ApJ SS, 154, 1-9
[2] Pilbratt G. L., Riedinger J. R., et al. 2010, A& A, 518, L1
[3] Holland W. S., Cunningham C. R., et al. 1998, Proc. SPIE Vol.3357, 305
[4] Ruhl J. E., Ade P. A. R., et al. 2004, Proc. SPIE Vol.5498, 11-29
[5] Condon J. J., 1992, ARA&A, 30, 575
[6] Murphy, E. J., Bremseth, J., Mason, B. S., et al. 2012, , 761, 97
[7] Padovani P., Miller N., Kellermann K. I., Mainieri V., Rosati P., Tozzi P., 2011, ApJ, 740, 20
[8] Cai, Z.-Y., Lapi, A., Bressan, A., et al. 2014, , 785, 65
[9] Granato, G. L., De Zotti, G., Silva, L., Bressan, A., & Danese, L. 2004, , 600, 580
[10] Lapi, A., Shankar, F., Mao, J., et al. 2006, , 650, 42
[11] Lapi, A., González-Nuevo, J., Fan, L., et al. 2011, , 742, 24
[12] Cai, Z.-Y., Lapi, A., Xia, J.-Q., et al. 2013, , 768, 21
[13] Bernardi, M., Shankar, F., Hyde, J. B., et al. 2010, , 404, 2087
[14] Mocanu, L. M., Crawford, T. M., Vieira, J. D., et al. 2013, ApJ, 779, 61
[15] Norris, R. P., Afonso, J., Bacon, D., et al. 2013, , 30, 20
[16] Clemens, M. S., Negrello, M., De Zotti, G., et al. 2013, , 433, 695
[17] Murphy, E. J., Condon, J. J., Schinnerer, E., et al. 2011, , 737, 67
[18] Draine, B. T. 2011, *Physics of the Interstellar and Intergalactic Medium* (Princeton University Press)

[19] Massardi M., Bonaldi A., Negrello M., Ricciardi S., Raccanelli A., De Zotti G., 2010, MNRAS, 404, 532

[20] Gardner J. P., Mather J. C., et al., 2006, Space Sci Rev, 123, 4, 485-606

[21] Mancuso C., Lapi A., Cai Z.Y., De Zotti G., Perrotta F., Negrello M., Danese L., 2014, in prep.

# Modeling the Mass Distribution in the Spiral Galaxy NGC 3198

**E. V. Karukes**[1,2], **P. Salucci**[1,2]

[1] SISSA/ISAS, International School for Advanced Studies, Via Bonomea 265, 34136, Trieste, Italy

[2]INFN, Sezione di Trieste, Via Valerio 2, 34127, Trieste, Italy

E-mail: ekarukes@sissa.it

**Abstract.** We study the HI Rotation Curve of the spiral galaxy NGC 3198 in terms of mass decomposition. We model the Rotation Curve in the framework of different models for the Dark Matter distribution: the Burkert profile and NFW profile. We show that Universal Rotation Curve (Burkert halo+stellar disk+gas disk) fits data accurately. Instead, the NFW (NFW halo+stellar disk+gas disk) model gives non-physical values of NFW halo parameters.

## 1. Introduction

The existence of significant discrepancies in galaxy clusters between the visible mass and the dynamical mass has been discovered by Zwicky in 1933 [1]. He measured the radial velocities for a set of galaxies in the Coma cluster, computed the velocity dispersion and concluded, by using the virial theorem, that for such large velocity dispersions the total mass should be 50 times greater than the one derived from luminous matter in galaxies .

Later, in the 1970s by Rubin, Ford and Kent [2] was delivered the first reliable evidence for Dark Matter (DM) in form of non-Keplerian Rotation Curve (RC). They analysed an optical RC of the Andromeda galaxy. They were expecting to see the declining RC with $V \propto \frac{1}{\sqrt{R}}$, where $V$ is the rotation velocity and $R$ is the galactocentric distance. The expected RC is called the Keplerian RC, as an analogy to Solar system. However, this Keplerian decline was not observed but it was obtained an approximately flat RC. Therefore, they concluded that galaxies should be surrounded by great quantities of non-luminous matter distributed far beyond the visible part of galaxies in more or less spherically symmetric form. This is consistent with the idea of the existence of massive DM halos proposed, at the same time, by Ostriker and Peebles [3] in order to stabilize the cold self-gravitating axisymmetric disks against bar-like models.

Today many authors consider that the inner regions of galaxies contain little or no DM. It means that the observed baryonic matter is responsible for the RC only in the inner region (e.g. [4, 5, 6, 7]). In the outer part, however, we must add the DM halo. Investigation of the outer part of galaxies became possible thanks to the interpretation of the HI rotation curves by Bosma in 1978 [8].

In brief, nowadays it is commonly accepted that RCs of spirals and also other type of galaxies [9, 10, 11, 12] can be interpreted as evidence of the existence of this DM component. The distribution of DM in galaxies can be described by a theoretical or an empirical density distribution profile. The most adopted theoretical model of the DM density profile, based on numerical simulations of the formation of galaxies in the Cold Dark Matter (CDM) scenario in the expanding Universe, is the NFW model proposed by Navarro, Frenk and White [13]. ΛCDM scenario describes well the large-scale structure of the Universe (e.g. [14]), but it seems to fail on the scales of galaxies and it is not confirmed by observed data (e.g. [15, 16, 17]). Observations seem to indicate an approximately constant dark matter density in the inner parts of galaxies, while ΛCDM scenario indicates a steep power-law-like behaviour. This difference has become known as the "core-cusp problem" to solve which, many other empirical profiles, such as the pseudo-isothermal (ISO) DM halo model [18, 19] and the Burkert profile [20] have been considered.

In this paper we study the RC of a spiral galaxy NGC 3198 in order to derive the properties of the DM halo around the galaxy. Our analysis is based on HI observations published by Gentile et al. [22]: these observations are part of HALOGAS (Westerbork Hydrogen Accretion in LOcal GAlaxieS) survey. In this paper they derived a very extended RC out to 720 arcsec (correspond to 48 kpc for a distance at 13.8 Mpc) and their mass distribution was modelled in the framework of modified Newtonian dynamics (MOND). In this paper we model the RC in the framework of the NFW and the Burkert profiles.

This paper is organized as follows. In Sec.2 we model the RC by using the quadrature sum of the RCs of the individual mass components (stellar disk + Dark halo + gas disk) where for Dark halo we apply first the Burkert density profile, and then the NFW profile. In Sec. 3 we provide the mass modeling results. Our conclusions are drawn in Sec. 4.

## 2. Mass Modelling

The data presented in the paper [22] allow us to make use of the most extended RC of a galaxy ever measured, in terms of optical disk scalelength. We model the spiral galaxy NGC 3198 as consisting of two "luminous" components, namely the stellar and gaseous disks, embedded in a dark halo.

### 2.1. Luminous matter

In order to study the properties of luminous and dark matter in this galaxy, we can decompose the observed RC into the stellar disk, the bulge, gas and the DM halo:

$$V^2(r) = V_{\rm d}^2(r) + V_{\rm b}^2(r) + V_{\rm g}^2(r) + V_{\rm DM}^2(r). \quad (1)$$

$V_{\rm d}(r)$ is the contribution of the stellar disk, was taken from [21]. Here, we use $V_{\rm d}^2 = \frac{\Upsilon_*^{3.6}}{0.57} V_{deBlok}^2$, where $V_{deBlok}^2$ are the data from [21] and the stellar mass-to-light ratio $\Upsilon_*^{3.6}$ is a free parameter (taking into account the scaling factor 0.57).

$V_{\rm b}(r)$ is the contribution from the bulge component. In NGC 3198 can be indicated an additional (third) compact, central component, which can be considered as a bulge. However, how it was found by [21], adding this contribution will not change the results, therefore we assume $V_{\rm bulge}(r) = 0$.

$V_{\rm g}(r)$ is the helium corrected contribution of the gaseous disk derived from the HI surface density distribution by [22].

$V_{\rm DM}(r)$ is the contribution of the dark matter halo which we discuss below.

### 2.2. Dark matter

For the DM halo with the virial mass $M_{vir} = \frac{4}{3}\pi 100\rho_{crit} r_{vir}^3$ several alternatives were investigated. In this paper we consider two of them.

Journal of Physics: Conference Series **566** (2014) 012008 doi:10.1088/1742-6596/566/1/012008

*2.2.1. Burkert halo* Burkert ([23, 20]) proposed an empirical density distribution of dark matter:

$$\rho_{\rm B}(r) = \frac{\rho_0 r_{\rm core}^3}{(r + r_{\rm core})(r^2 + r_{\rm core}^2)}, \tag{2}$$

where $\rho_0$ (the central density) and $r_{\rm core}$ (the core radius) are the two free parameters.

*2.2.2. Navarro, Frenk and White (NFW) halo* Navarro et al. in 1996 [13] proposed a profile based on numerical simulations of the formation of galaxies in the CDM scenario. The NFW profile is written as

$$\rho_{\rm NFW}(r) = \frac{\rho_{\rm s}}{(\frac{r}{r_s})(1 + \frac{r}{r_s})^2}, \tag{3}$$

where $\rho_{\rm s}$ and $r_s$ are respectively the characteristic density and the scale radius of the distribution. These two parameters are in principle independent, but they can also be expressed in terms of the virial mass $M_{\rm vir}$, the concentration parameter $c = \frac{r_{\rm vir}}{r_s}$ and the critical density of the Universe $\rho_{\rm crit} = 9.31 \times 10^{-30} {\rm g/cm}^3$ [24, 25]. By using Eq.(3), we can write:

$$\begin{aligned} \rho_s &= \frac{100}{3} \frac{c^3}{\log(1+c) - \frac{c}{1+c}} \rho_{\rm crit} \quad {\rm g/cm}^3; \\ r_s &= \frac{1}{c} \left( \frac{3 \times M_{\rm vir}}{4\pi 100 \rho_{\rm crit}} \right)^{1/3} \quad {\rm kpc}. \end{aligned} \tag{4}$$

The relation between $c - M_{\rm vir}$ is taken from [26]:

$$c \simeq 9.60 \left( \frac{M_{\rm vir}}{10^{12}(0.71)^{-1} M_\odot} \right)^{-0.075}. \tag{5}$$

## 3. Results

The results of mass decompositions are shown in Fig. 1 and Fig. 2.

We start from the results for the Burkert halo which gives a very good quality fit with a reduced chi-square value of $\chi^2 = 1.6$. The best-fit parameters of the Burkert halo are

$$\begin{aligned} \rho_0 &= (3.58 \pm 0.73) \times 10^{-25} \quad {\rm g/cm}^3; \\ r_{\rm c} &= (16.51 \pm 2.0) \quad {\rm kpc}; \\ \Upsilon_*^{3.6} &= (0.97 \pm 0.05). \end{aligned}$$

Then the dynamical mass of the stellar disk is

$$M_{\rm D} \simeq 1.1 \frac{\Upsilon_*^{3.6}}{0.57} V_{deBlok}^2(R_{\rm l}) \frac{R_{\rm l}}{G} \tag{6}$$

where $V_{deBlok}$ is the contribution to the circular velocity of the stellar disk at a very extended point $R_{\rm l} \approx 48.2$ kpc. We found $M_{\rm D} \simeq 4.5 \times 10^{10} M_\odot$ with a propagated uncertainty of about 10 %. The corresponding virial mass and radius of the DM halo are $M_{\rm vir} = (5.7^{+1.3}_{-0.6}) \times 10^{11} M_\odot$ and $R_{\rm vir} = (212^{+11}_{-7})$ kpc as indicated above.

Journal of Physics: Conference Series **566** (2014) 012008 doi:10.1088/1742-6596/566/1/012008

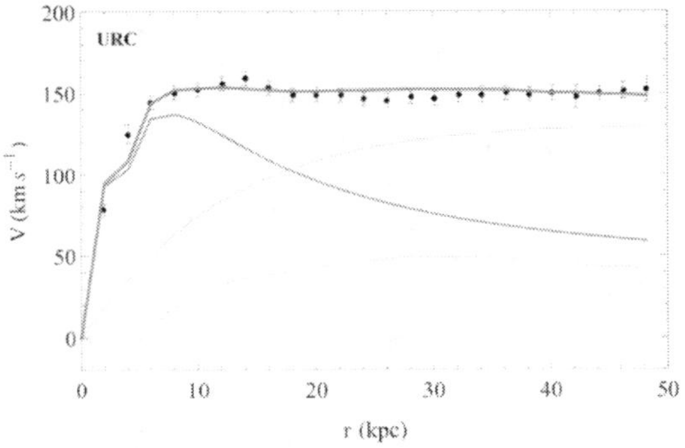


**Figure 1.** Burkert model of NGC 3198. The circular velocity data (filled circles with error bars) is modeled (thick Red line) with the Burkert profile (thick Green line), the disk (Magenta line) and the HI circular velocity (Azure line).

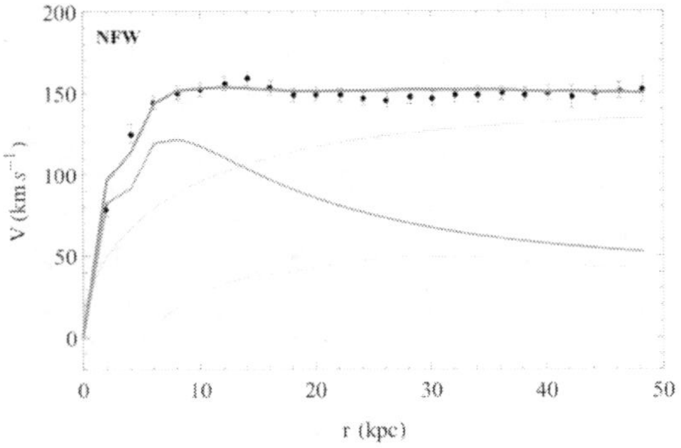


**Figure 2.** NFW model of NGC 3198. The circular velocity data (filled circles with error bars) is modeled (thick Red line) including the disk (Magenta line), the NFW halo profile (Green line) and the HI circular velocity (Azure line).

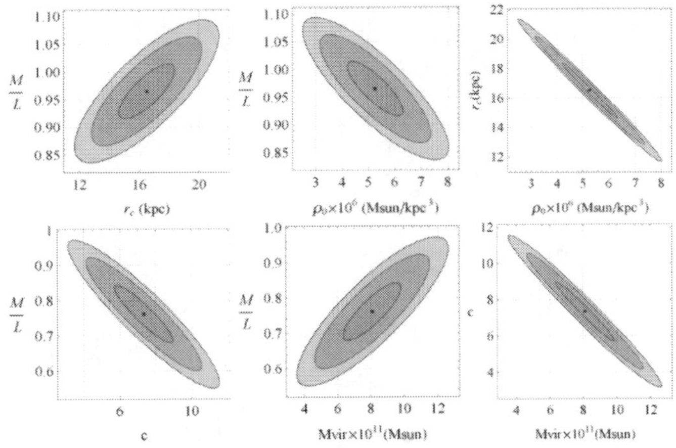


**Figure 3.** Top panel: 1, 2, 3$\sigma$ confidence ellipses (Purple, Red, Orange respectively) for the best-fit parameters in the Burkert halo case. The central points indicate the best-fitting values, $r_c$ is in units $kpc$, $\rho_0$ is in units $M_\odot/kpc^3$.
Bottom panel: 1, 2, 3$\sigma$ onfidence ellipses (Purple, Red, Orange respectively) for the best-fit parameters in the NFW halo case. The central points, as well, indicate the best-fitting values, $M_{\rm vir}$ is in units $10^{11}M_\odot$.

For the NFW profile we fitted data in terms of the virial mass, the concentration parameter and the mass-to-light ratio ($M_{\rm vir}, c, \Upsilon_*^{3.6}$). In this case we have got the same value of a reduced chi-square, even slightly better, $\chi^2 = 1.5$ . However, the value of the concentration parameter is inconsistent with the one found in Eq.(5), where $c \approx 10$. This discrepancy was found also in other works [27, 28].

In detail the results from the best-fit are

$$M_{\rm vir} \;=\; (8.1 \pm 1.8) \times 10^{11} \quad {\rm M}_\odot;$$

Journal of Physics: Conference Series **566** (2014) 012008 doi:10.1088/1742-6596/566/1/012008

$$c = (7.3 \pm 1.6);$$
$$\Upsilon_*^{3.6} = (0.80 \pm 0.10).$$

The corresponding values of the scale radius and the characteristic density are $r_s = 32.9 \pm 9.7$ kpc and $\rho_s = (9.9 \pm 5) \times 10^{-26}$ g/cm$^3$. The too high value of the scale radius $r_s$ and the too low value of the characteristic density $\rho_s$ are unrealistic.

The 1,2,3-$\sigma$ confidence regions for the more relevant best-fit parameters are shown in Fig. 3. The central points correspond to the best-fit values.

## 4. Conclusions

We have used new HI observations of the spiral galaxy NGC 3198, performed by the HALOGAS survey, for the mass modeling. The HI disk of this galaxy is very extended, up to 13 disk length scale. The derived RC is dominated by DM at large galactocentric radii, where baryons are not able to account for the observed kinematics.

In this work we have used two DM profiles. The (cored) URC Burkert dark halo and NFW halo fit the data very well. However, the best-fit paremeters in NFW case are found outside the range expected in the ΛCDM scenario.

## 5. References


[1] Zwicky F 1933 *Helv. Phys. Acta* **6** 110
[2] Rubin V C, Ford W and Kent Jr 1970 *J. ApJ.* **159** 379
[3] Ostriker J P and Peebles P J E 1973 *ApJ* **186** 467
[4] van Albada T S, Bahcall J N, Begeman K and Sancisi R 1985 *AJ* **295** 305
[5] Athanassoula E, Bosma A and Papaioannou S 1987 *A&A* **7279** 23
[6] Persic M and Salucci P 1988 *MNRAS* **234** 131
[7] Palunas P and Williams T B 2000 *AJ* **120** 2884
[8] Bosma A 1978 *PhD Thesis* **195**
[9] Carignan C 1985 *ApJ* **299** 59
[10] Swaters R 1999 *PhD Thesis, University of Groningen*
[11] Danziger I J 1997, in Persic M and Salucci P (Eds.) *Dark and visible matter in galaxies*, ASP Conf. Series **117** 28
[12] Frusciante N, Salucci P, Vernieri D, Cannon J M and Elson E C 2012 *MNRAS* **426** 751
[13] Navarro J F, Frenk C S and White S D M 1996 *ApJ* **462** 563
[14] Springel C S, Frenk C S and White S D M 2006 *Nature* **440** 1137
[15] de Blok W J G and Bosma A 2002 *A&A* **385** 816
[16] Gentile G, Salucci P, Klein U, Vergani D and Kalberla P 2004 *ApJ* **634** 2,L145
[17] Gentile G, Burkert A, Salucci P, Klein U and Walter F 2005 *MNRAS* **351** 903
[18] Begeman K G 1989 *A&A* **223** 47
[19] Kent S M 1986 *AJ* **91** 1301
[20] Salucci P and Burkert A 2000 *AJ* **537** L9,L12
[21] de Blok W J G, Walter F, Brinks E, Trachternach C, Oh S-H and Kennicutt R C Jr 2008 *AJ* **136** 6,2648
[22] Gentile G,Jozsa G I G, Serra P, Heald G H, de Blok W J G, Graternali F, Patterson M T, Walterbos R A M and Osterloo T 2013 *A&A* **554** A125
[23] Burkert A 1995 *ApJL* **447** L25
[24] Bullock J S, Kolatt T S, Rachel Y S, Somerville S, Kravtsov A V, Klypin A A, Primsck J R and Dekel A 2001 *MNRAS* **321** 559
[25] Wechsler R H, Bullock J S, Primack J L, Kravtsov A V and Dekel A 2002 *A&A* **66** 505
[26] Klypin A A, Trujillo-Gomez S and Primack J 2011 *ApJ* **740** 102
[27] Memola E, Salucci P and Babic A 2011 *A&A* **534** 50
[28] Nesti F and Salucci P 2013 *JCAP* **07** 016

Journal of Physics: Conference Series **566** (2014) 012009 doi:10.1088/1742-6596/566/1/012009

# The intrinsic shape and dynamical structure of the bulges of lenticular galaxies

**L Costantin[1], E M Corsini[1,2], J Méndez-Abreu[3], E Dalla Bontà[1,2], L Morelli[1,2] and A Pizzella[1,2]**

[1] Dipartimento di Fisica e Astronomia, Università di Padova, Italy
[2] INAF-Osservatorio Astronomico di Padova, Italy
[3] School of Physics and Astronomy, University of St. Andrews, UK

E-mail: luca.costantin@studenti.unipd.it, enricomaria.corsini@unipd.it, jma20@st-andrews.ac.uk, elena.dallabonta@unipd.it, lorenzo.morelli@unipd.it, alessandro.pizzella@unipd.it

**Abstract.** Several scenarios have been proposed to account for the formation and evolution of galactic bulges and to explain the variety of their observed properties. Both the intrinsic shape and dynamical structure of bulges depend on the mechanisms and timescales of their formation. We are deriving the intrinsic axial ratios of the bulges of a sample of unbarred lenticular galaxies to look for a possible relationship with their known orbital structure. Preliminary results show that the orbits of stars of the triaxial bulge of NGC 4476 are more anisotropic than those in the axisymmetric bulge of NGC 4249 hinting at a different formation process.

## 1. Introduction

Several formation scenarios were proposed to explain the variety of photometric, kinematic, dynamical, and stellar population properties observed in bulges of lenticular and spiral galaxies. Classical bulges are similar to low-luminosity elliptical galaxies and they are thought to be formed by rapid gravitational collapse or hierarchical merging; boxy/peanut-shaped bulges are bars and formed from dynamical instabilities of the galactic disks; pseudobulges are reminiscent of disks and formed from secular processes of redistribution of the gas and stars driven by bars or environmental effects (see [1], for a review). This picture is further complicated by the co-existence of mixed types of classical and pseudobulges in the same galaxy [2, 3] and their combination with boxy/peanut structures in barred galaxies [4, 5].

Studying the intrinsic shape of the bulges provides a crucial piece of information for testing the results of numerical simulations of bulge formation. About 80% of bulges in unbarred lenticular and early-to-intermediate spiral galaxies are not oblate but triaxial ellipsoids [6] and the distribution of bulge triaxiality is strongly bimodal [7]. According to numerical simulations, bulges with a Sérsic index $n \leq 2$ which show a high fraction of oblate axisymmetric (or nearly axisymmetric) shapes and have a bulge-to-total ratio $B/T \leq 0.3$, may be the result of dissipational minor mergers. Both major dissipational and dissipationless mergers seem to be required to explain the variety of shapes found for bulges with $n > 2$ and $B/T > 0.3$ (e.g., [8]).

We started a project to analyze the structural and dynamical properties of the bulges of a sample of unbarred lenticular galaxies in order to study the relationship between their intrinsic

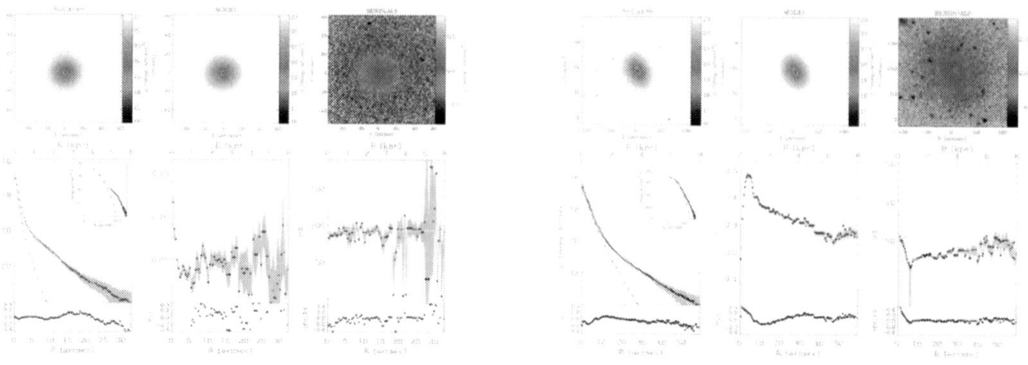

**Figure 1.** Two-dimensional photometric decomposition of the fast-rotating galaxy NGC 4249 (left plot) and slow-rotating galaxy NGC 4476 (right plot) obtained from GASP2D. Upper panels (from left to right): map of the observed, modeled, and residual (observed−modeled) surface brightness distribution of the galaxy. Lower panels (from left to right): ellipse-averaged radial profile of surface brightness, position angle, and ellipticity measured in the observed (dots with error bars) and modeled image (green solid line). The surface-brightness radial profiles of the fitted bulge (blue dashed line) and disk (red dotted line) are also shown.

shape and orbital structure. In this paper we present the preliminary results for the bulges of the two S0 galaxies NGC 4249 and NGC 4476.

## 2. Galaxy sample

Our sample comprises 46 objects from the volume-limited ATLAS$^{3D}$ sample of 260 early-type galaxies with a distance $d < 40$ Mpc [9]. The sample galaxies were selected to be classified as SA0 [10]; with inclination $i < 65°$ to successfully perform the photometric decomposition; without morphological (e.g., signatures of interactions; [9]) and/or kinematic (e.g., evidence for decoupled components; [11]) peculiarities; with $i$-band images in the data archive of the Sloan Digital Sky Survey (SDSS), [12].

## 3. Photometric parameters of the bulge and disk

The photometric decomposition of the $i$-band SDSS images of the sample galaxies was performed using the GASP2D algorithm [6], which yields the structural parameters for a Sérsic bulge and an exponential disk.

The fitting algorithm relies on a $\chi^2$ minimization of the intensities in counts, for which we must adopt initial trial parameters that are as close as possible to their final values. The trial values of the effective surface brightness, effective radius, shape parameter, position angle, and axial ratio of the bulge ($I_{\rm e}$, $r_{\rm e}$, $n$, PA$_{\rm b}$, and $q_{\rm b}$) and of the central surface brightness, scale-length, position angle, and axial ratio of the disk ($I_0$, $h$, PA$_{\rm d}$, and $q_{\rm d}$) were estimated from the analysis of the ellipse-averaged radial profiles of surface brightness $\mu_i$, ellipticity $\epsilon$, and position angle PA. The latter were measured using the IRAF task ELLIPSE [13]. Starting from these initial trial parameters the different photometric models of the surface brightness were fitted iteratively to the galaxy image. Each image pixel was weighted according to the variance of its total observed photon counts due to the contribution of both galaxy and sky, and determined assuming photon noise limitation and taking the detector read-out noise into account. Seeing effects were also

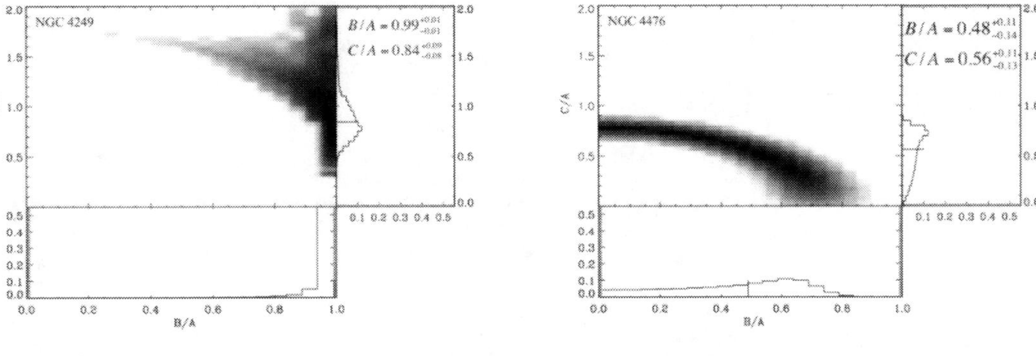


**Figure 2.** The distribution of the intrinsic axial ratios of the bulge of NGC 4249 (left plot) and NGC 4476 (right plot). For each galaxy, the median value of $B/A$ and $C/A$ is given and marked with a segment in the bottom and right panel, respectively.

taken into account by convolving the model image with a circular Moffat point spread function (PSF) with a FWHM matching the observed one. The convolution was performed as a product in Fourier domain before the least-squares minimization. Only the image pixels with an intensity larger than the sky standard deviation were included in the fit. Foreground stars were masked and excluded from the fit.

The analysis of the fitted images allowed to reject 29 galaxies since they either turned out to be ellipticals or showed a bar and/or spiral arms. The result of the photometric decomposition of the images of NGC 4249 and NGC 4476 is shown in figure 1.

The errors on the fitted parameters were estimated through Monte Carlo simulations on images of artificial galaxies. Finally, a background level and photon noise were added to the simulated images in order to mimic the instrumental setup and image signal-to-noise ratio. The relative errors in the fitted parameters of the artificial galaxies were estimated by comparing the input and output values and were assumed to be normally distributed. The standard deviation was adopted as the $1\sigma$ error in the relevant parameter for the bulge-disk decomposition. Systematic errors given by a wrong estimation of the PSF FWHM and sky level are the most significant contributors to the error budget. To derive the intrinsic shape of the bulge we are interested in $q_{\rm b}$, $q_{\rm d}$, $\rm PA_b$, and $\rm PA_d$. Their $3\sigma$ errors including systematics are smaller than 15%.

## 4. Intrinsic shape of the bulges

The intrinsic shape of the sample bulges was derived from the geometrical relationships between the projected and intrinsic shapes of the bulge and its surrounding disk (see [7] for details). The bulge is assumed to be a triaxial ellipsoid with semi-axes of length $A$ and $B$ in the equatorial plane and $C$ along the polar axis. The bulge shares the same center and polar axis as the disk, which is circular and lies on the equatorial plane of the bulge. The intrinsic shape of the bulge is recovered from the bulge ellipticity parameter $e = (1 - q_{\rm b}^2)/(1 + q_{\rm b}^2)$, twist angle $\delta = 180° - |\rm PA_b - PA_d|$ between the major axes of the bulge and disk (i.e., the line of nodes), and disk inclination $\theta = \arccos q_{\rm d}$.

The relation between the intrinsic and projected parameters depends only on the spatial position of the bulge, i.e., on the angle $0 \leq \phi \leq 90°$ measured on the bulge equatorial plane between the principal axis corresponding to $A$ and the line of nodes. The equatorial ellipticity $Z = B^2/A^2$ and intrinsic flattening $F = C^2/A^2$ of the bulge are given by:

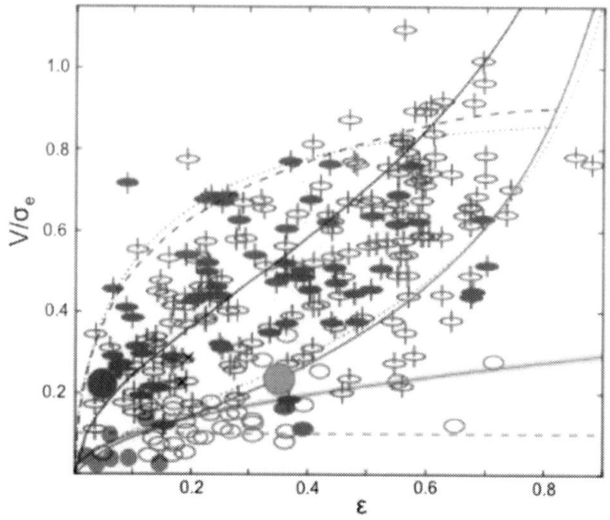


**Figure 3.** ($\epsilon$, $V/\sigma$) diagram of the ATLAS$^{3D}$ galaxies (see [14] for details). NGC 4249 (blue circle) and NGC 4476 (orange circle) are shown together with the non-rotating galaxies (red symbols), non-regular rotators without any specific kinematic feature (green symbols), and regular rotators (purple symbols). The solid black line corresponds to isotropic oblate systems viewed edge-on.

$$\frac{\sin(2\phi_C)\sin^2\theta}{\cos^2\theta}F = -\sin\phi_B\cos(2\phi_C-\phi_B)(1+Z)^2 + \sin(2\phi_C-\phi_B)\sqrt{(1-Z)^2-\sin^2\phi_B(1+Z)^2}\,, \tag{1}$$

where:

$$\phi_B = \arctan\frac{e\sin 2\delta}{\cos\theta\,(1+e\cos 2\delta)}\,, \tag{2}$$

$$\phi_C = \frac{1}{2}\arctan\frac{2\,e\sin 2\delta\,\cos\theta}{e\cos 2\delta\,(1+\cos^2\theta)-\sin^2\theta}\,. \tag{3}$$

Since the axial ratios $B/A$ and $C/A$ are both functions of the same variable $\phi$, their probabilities are identical, i.e., for a given value of $B/A$ with probability $P(B/A)$, the corresponding value of $C/A$ obtained by equation (1) has a probability $P(C/A) = P(B/A)$. This allows to obtain the range of possible values of $B/A$ and $C/A$ for the bulge and to constrain its most probable intrinsic shape by adopting the probabilities $P(Z)$ and $P(F)$, [7].

The distribution of $B/A$ and $C/A$ for the bulges of NGC 4249 and NGC 4476 as calculated via Monte Carlo simulations is shown in figure 2. We randomly chose 5000 geometric configurations assuming a Gaussian distribution of $q_{\rm b}$, $q_{\rm d}$, $\mathrm{PA_b}$, and $\mathrm{PA_d}$ taking into account their $3\sigma$ errors. For each geometric configuration we derived 1000 values of $B/A$ and $C/A$ according to their probability distribution functions. The resulting median values of the axial ratios and the $1\sigma$

confidence intervals from their cumulative distribution are $B/A = 0.99^{+0.01}_{-0.01}$ and $C/A = 0.84^{+0.09}_{-0.08}$ for the bulge of NGC 4249 and $B/A = 0.48^{+0.11}_{-0.14}$ and $C/A = 0.56^{+0.11}_{-0.13}$ for the bulge of NGC 4476, respectively.

## 5. Conclusions

The two S0 galaxies NGC 4249 and NGC 4476 are classified as a fast and slow rotator, respectively (figure 3, [14]). Slow rotators exhibit complex stellar velocity fields and often include stellar kinematically distinct cores, fast rotators have regular velocity fields and show aligned photometric and kinematic axes.

We have found that the bulge of NGC 4249 is axisymmetric and almost spherical, whereas the bulge of NGC 4476 is triaxial and partially prolate (figure 2). From the location of NGC 4249 and NGC 4476 in the ($\epsilon$, $V/\sigma$) diagram (figure 3), we conclude that the orbits of stars of the triaxial bulge of NGC 4476 are more anisotropic than those in the axisymmetric bulge of NGC 4249.

This is in agreement with the recent findings on the shape of the ATLAS$^{3D}$ early-type galaxies [15]. It has been shown that the fast rotators are predominantly axisymmetric, whereas slow rotators are mildly triaxial hinting at different formation scenarios. However, their photometric and kinematic properties are representative of the galaxy as a whole, and therefore, no attempt to account for the different structural components of the galaxies (i.e., bulge, disk and bar) or even to distinguish elliptical and lenticular galaxies was made. Moreover, a Gaussian intrinsic shape distribution under assumption of axisymmetry has been adopted in the photometric analysis. We plan to address the above issues for the bulges of lenticular galaxies by completing the analysis of their intrinsic shape and orbital structure for all the objects of our sample.

## Acknowledgments

This work was supported by Padua University through grants 60A02–5052/11, 60A02–4807/12, 60A02–5857/13, and CPDA133894. J.M.A. acknowledges support from the European Research Council Starting Grant (SEDmorph; P.I. V. Wild). L.M. received financial support from Padua University grant CPS0204. L.C. acknowledges the University of St. Andrews for their hospitality while this paper was in progress.

## References

[1] Kormendy J and Kennicutt R C 2004 *ARA&A* **42** 603
[2] Gadotti D A 2009 *MNRAS* **393** 1531
[3] Nowak N, Thomas J, Erwin P, Saglia R P, Bender R and Davies R I 2010 *MNRAS* **403** 646
[4] Athanassoula E 2005 *MNRAS* **358** 1477
[5] Méndez-Abreu J, Debattista V P, Corsini E M and Aguerri J A L 2014 *A&A* in press
[6] Méndez-Abreu J, Aguerri J A L, Corsini E M and Simonneau E 2008 *A&A* **478** 353
[7] Méndez-Abreu J, Simonneau E, Aguerri J A L and Corsini E M 2010 *A&A* **521** A71
[8] Cox T J, Jonsson P, Primack J R and Somerville R S 2006 *MNRAS* **373** 1013
[9] Cappellari M et al. 2011 *MNRAS* **413** 813
[10] de Vaucouleurs G, de Vaucouleurs A, Corwin H G, Jr, Buta R J, Paturel G and Fouqué P 1991 *Third Reference Catalogue of Bright Galaxies* (New York, NY: Springer)
[11] Krajnović D et al. 2011 *MNRAS* **414** 2923
[12] York D G et al. 2000 *AJ* **120** 1579
[13] Jedrzejewski R I 1987 *MNRAS* **226** 747
[14] Emsellem E et al. 2011 *MNRAS* **414** 888
[15] Weijmans A-M et al. 2014 *MNRAS* in press

Journal of Physics: Conference Series **566** (2014) 012010 doi:10.1088/1742-6596/566/1/012010

# Decoding the stellar fossils of the dusty Milky Way progenitors

**M de Bennassuti[1,2], R Schneider[2], R Valiante[2] and S Salvadori[3]**

[1] Dipartimento di Fisica, Sapienza, Università di Roma, Piazzale Aldo Moro 5, 00185 Roma, Italy
[2] INAF - Osservatorio Astronomico di Roma, Via di Frascati 33, 00040 Monte Porzio Catone, Italy
[3] Kapteyn Astronomical Institute, University of Groningen, Landleven 12, 9747 AD Groningen, The Netherlands

E-mail: matteo.debennassuti@oa-roma.inaf.it

**Abstract.** We investigate the metallicity distribution function (MDF) of the Galactic halo and the relative fraction of Carbon-normal and Carbon-rich stars using the semi-analytical code GAMETE. The code reconstructs the hierarchical merger tree of the Milky Way (MW) and follows the star formation history and the metal evolution in individual progenitors, including for the first time the formation and evolution of dust. We predict scaling relations between the dust, metal and gas masses for MW progenitors and compare them with observational data of galaxies at $0 \leq z < 6.3$. We find that the relative contribution of C-normal and C-enhanced stars to the MDF and its dependence on [Fe/H] allow to discriminate among different Pop III/II transition criteria as well as between different Initial Mass Functions (IMFs) and supernova (SN) yields for Population III stars.

## 1. Introduction

The physical conditions which enable the formation of the first low-mass and long-lived stars to form in the Universe is still a subject of debate. Numerical studies suggest the first Population III (Pop III) stars have masses mostly distributed between a few tens and a few hundred solar masses. The first supernovae (SNe) explode and seed the gas with metals and dust grains, changing the physical properties of star forming regions, eventually allowing the formation of the first low-mass Population II (Pop II) stars. Gas at low metallicity can achieve larger cooling rates through additional molecular species (HD, OH, CO, $H_2O$), fine structure line–cooling (mostly OI and CII) and thermal emission from dust grains. The relative importance of these coolants depends on the density (time) regime during the collapse and on the metal/dust content of the collapsing core.

The transition between Pop III and Pop II may be guided by different critical parameters: metallicity, $Z_{cr} = 10^{-3.8} Z_\odot$ [1], dust-to-gas mass ratio, $\mathcal{D}_{cr} \sim 4.4 \times 10^{-9}$ [2] or transition discriminant, $D_{trans,cr} = -3.5 \pm 0.2$ [3]. Most of the stars observed in the Galactic halo or in its dwarf satellites have $D_{trans} = \log_{10}(10^{[C/H]} + 0.9 \times 10^{[O/H]})$[1] $> D_{trans,cr}$, with the exception of SDSS J102915 [5]; this star provides a strong indication that low-mass star formation may have occurred in the absence of efficient metal-line cooling, supporting a dust-driven transition [6].

Hence, stellar archaeology of the most metal-poor stars is a promising way to constrain the environmental conditions for the formation of the first low-mass stars, exploring the star formation and the chemical enrichment of the Universe. In particular, the low-metallicity tail of the metallicity

[1] $[X/Y] = \log_{10}(N_X/N_Y) - \log_{10}(N_X/N_Y)_\odot$, for elements X and Y. Solar value are taken from [4].

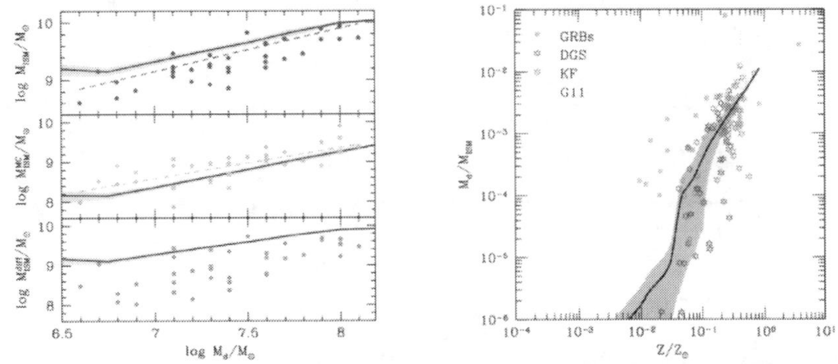

**Figure 1.** Comparison between the predicted properties of MW progenitors and observations. Solid lines represent averages over 50 independent merger histories of the MW and shaded regions the 1-$\sigma$ dispersion. *Left panel*: correlation between the dust mass and the diffuse, molecular and total gas masses (from bottom to top). Observational data are taken from [7] with dashed lines indicating the best-fit linear relation for the data. *Right panel*: dust-to-gas mass ratio as a function of the gas metallicity. Stars refer to the sample analyzed by [8] and squares indicate the GRB data sample of [9].

distribution function (MDF, i.e. the number distribution of stars as a function of their [Fe/H]) at [Fe/H] $< -2$ and the observed surface elemental abundances of the most metal-poor stars can provide observational constraints on the nature of Pop III stars and on the formation efficiency of the first Pop II stars. Other hints are obtained by the so-called Carbon-enhanced extremely metal-poor stars (CEMP, with [C/Fe] $> +1$), whose fraction grows with decreasing [Fe/H] [10].

Here we describe an improved version of the hierarchical semi-analytical code GALAXY MERGER TREE AND EVOLUTION (GAMETE) [11]. Using this model, we predict the stellar, gas, metal and dust content of all the MW progenitors along the simulated merger trees, comparing our predictions with observations of the global properties of the MW and with the observed scaling relations inferred by different galaxy samples. We then apply the model to stellar archaeology to constrain the nature of the first stars.

## 2. Description of the model

The GAMETE code allows to reproduce the observed global properties of the MW, such as the mass of gas and stars, metallicity and dust-to-gas ratio, predicting their evolution along cosmic time. Enrichment processes are regulated by both Asymptotic Giant Branch (AGB) stars and SNe, which eject metals and dust into the ISM according to their stellar lifetime [12], accounting for partial dust destruction by the reverse shock in SN ejecta. We follow the formation and evolution of a MW-like galaxy in the framework of the Lambda Cold Dark Matter ($\Lambda$CDM) model using a binary Monte Carlo algorithm based on the Extended-Press and Schechter theory. The code traces the merger history backward in time, starting from a MW-like DM halo of $M_{MW} = 10^{12} M_{\odot}$ at redshift $z = 0$ up to $z \sim 20$.

We have further improved GAMETE to describe the evolution of the gas in two separate phases of the ISM: a *diffuse* component (warm/hot low-density gas), where dust can be destroyed by SN shocks, and a *dense* or *molecular cloud* (MC) component (cold and dense gas), where star formation occurs and where dust grains can grow in mass by accreting gas-phase metals, being shielded from destructive processes.

Pop III stars are assumed to form with a Larson IMF (a power-law with an exponential cut-off below a characteristic mass, $m_{ch}$, chosen to have a $\sim 40\,M_{\odot}$ average mass [13]) either with masses in the range $[10-300]\,M_{\odot}$ or with masses in the range $[10-140]\,M_{\odot}$. Pop III stars enrich the ISM with metal and dust yields appropriate for core-collapse SNe [14, 15] or faint SNe [16] ($10\,M_{\odot} < m_* < 40\,M_{\odot}$) and PISNe [17, 18] ($140\,M_{\odot} < m_* < 260\,M_{\odot}$); outside these two mass ranges stars are assumed to directly collapse

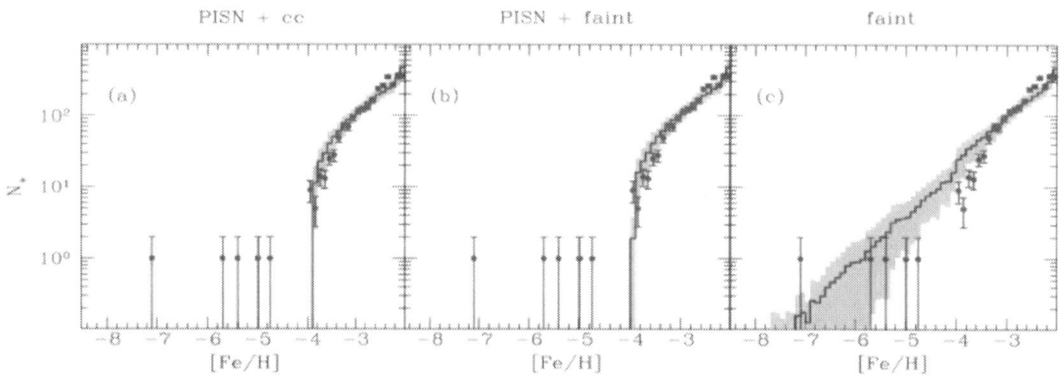


**Figure 2.** Simulated MDF as a function of [Fe/H] abundance. Solid lines show the results averaged over 50 merger trees with the shaded areas representing the 1-$\sigma$ dispersion. The three panels are for: (*a*) PopIII IMF extending up to 300 $M_\odot$ with PISNe and core-collapse SN yields; (*b*) PopIII IMF up to 300 $M_\odot$ with PISNe and faint SNe yields; (*c*) PopIII IMF with stellar masses up to 140 $M_\odot$ and faint SNe yields only. Here we show the results obtained adopting the critical metallicity $Z_{cr} = 10^{-3.8} Z_\odot$ as transition criterium. Other criteria provide very similar results.

to black hole without ejecting the products of their nucleosynthesis. The main feature of faint SNe is the dramatically lower Fe yield (~ 5 orders of magnitude smaller with respect to ordinary core-collapse SNe).

Once the criterium for low-mass star formation is met (see Sec. 1), Pop II stars form according to a Larson IMF ([0.1 - 100] $M_\odot$, $m_{ch} = 0.35\ M_\odot$). Pop II yields account for both AGB [19, 20] and SNe [14, 15] contributions for stars with $m_* < 40\, M_\odot$.

The model free parameters are calibrated so as to reproduce the MW observations on the total mass of stars, the SFR and the mass of gas, dust and metals in the two ISM phases. Hereafter we refer to our fiducial model as the model where the Pop III/II transition occurs when $\mathcal{D} \geq \mathcal{D}_{cr} = 4.4 \times 10^{-9}$ and Pop III stars form in the range [10 – 140] $M_\odot$. All the details of this study are presented in [21].

## 3. Properties of the Milky Way progenitors

In this work we analyse the properties of the MW progenitors predicted by our reference model. By selecting the progenitors with stellar, dust and gas masses comparable to that of the observed sample of galaxies we have tested the predicted dust-to-metals and dust-to-gas scaling relations of our model.

In the left panel of Fig. 1 we compare the model predictions with local galaxies [7]. The behaviour of the simulated progenitors is globally consistent with observed data, with a slope close to the proposed linear regression (dashed lines). The correlation between the dust mass and total gas mass shows that MW progenitors are generally more gas-rich than typical galaxies in the observed sample, probably as a consequence of the earlier evolutionary stages of the model galaxies, which lie in the redshift range $0 < z < 4$. It is important to note that galaxies in the [7] sample have been classified to be HI-deficient, probably due to galaxy interactions in the cluster environment where they are embedded.

Additional constraints on the model can be obtained by comparing the predicted correlation between the dust-to-gas ratio and the gas metallicity. In the right panel of Fig. 1, we show the data points from three local galaxy surveys collected by [8] and the data collected by [9] over a wide redshift range $0.1 < z < 6.3$ using GRB afterglows. The observed data points indicate that the dust-to-gas ratio grows with the metallicity but with a large scatter, particularly at the lowest metallicities. A similar behaviour is found for the simulated MW progenitors.

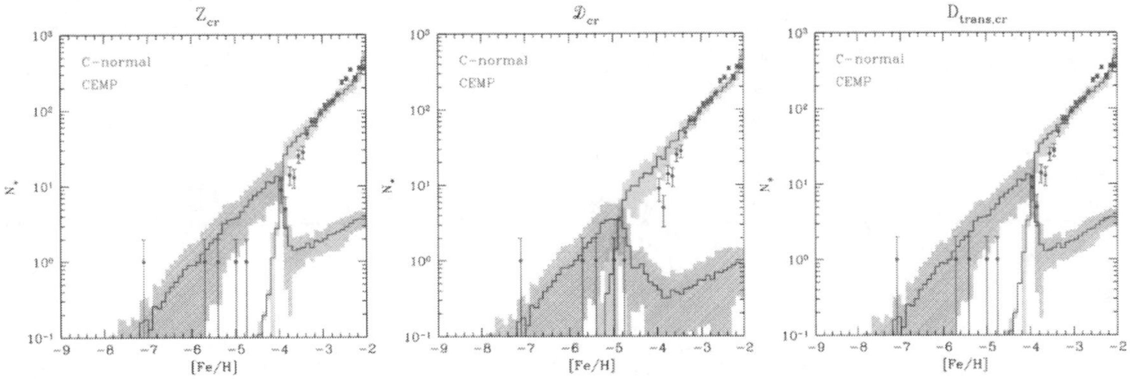


**Figure 3.** The contribution of C-enhanced stars ( [C/Fe] $\geq$ 1.0, purple line and shaded region) and C-normal stars ([C/Fe] < 1, blue line and shaded region) to the theoretical MDF for different transition criteria ($Z_{cr}$, $\mathcal{D}_{cr}$, $D_{trans,cr}$, from left to right).

## 4. Stellar Archaeology

Here we study the predicted halo MDF for different Pop III mass ranges and/or Pop III SN yields, varying the criterium adopted for the Pop III/II transition.

**Constraints on the Pop III IMF.** Fig. 2 shows the comparison of models where the Pop III/II transition is driven by a critical metallicity of $Z_{cr} = 10^{-3.8} Z_{\odot}$. They differ only for the adopted Pop III IMF and corresponding yields, while the data points refer to the joint HK/HES sample that has already been used in [11]. We have added the 5 hyper-iron-poor stars currently known with [Fe/H] < −4.5.

The Pop III IMF, from left to right, extends up to (*a*) 300 $M_{\odot}$ (yields from cc-Sne and PISNe), (*b*) 300 $M_{\odot}$ (yields from faint SNe and PISNe) and (*c*) 140 $M_{\odot}$ (yields from faint SNe). In panels (*a*) and (*b*) it is evident that PISNe dominate the enrichment over cc and faint SNe, leading to the formation of stars with [Fe/H] $\gtrsim$ -4, at odds with observations. Panel (*c*) shows that if PISNe are excluded, then the iron enrichment is much slower, being dominated by faint SNe. As a result, the tail of the MDF extends to [Fe/H] ~ -8, well into the observed range of the most iron-poor stars. The excess of predicted stars with −4.75 < [Fe/H] < −3.5 suggests that additional physical processes are at work, as radiative feedback effects that may prevent the gas in the first mini-halos from cooling, delaying or inhibiting star formation.

**Constraints on the Pop III/II transition.** For the same set of Pop III IMF and SN yields discussed above, the predicted MDF for different transition criteria are very similar to the $Z_{cr}$ case: the mass range of PISN progenitors has to be excluded and faint SNe are needed to reproduce the low-metallicity tail of the MDF for all transition criteria, but the full MDF cannot discriminate between different transition criteria, being very similar in the three cases.

Tighter constraints on the Pop III/II transition can be set by separating the contribution to the MDF of C-normal and C-enhanced stars. The results are shown in Fig. 3. It is clear that CEMP stars dominate the low-[Fe/H] tail of the simulated MDF for all three cases. However, it is only in the case where Pop II stars from when $\mathcal{D} \geq \mathcal{D}_{crit}$, that the MDF of C-normal stars extends to [Fe/H] ~ −5.4, accounting for the observed data point that corresponds to SDSS J102915, the only C-normal star at these low-[Fe/H] currently known. Hence, the formation of SDSS J102915 must have followed alternative pathways, beyond that associated to fine-structure line cooling.

**Carbon enhancement.** Spectroscopic studies of metal poor stars have convincingly shown that the frequency of CEMP stars grows with decreasing [Fe/H] [10]. In Fig. 4 (right panel) we plot the observed fraction of CEMP stars (number of CEMP stars divided by the total number of stars) and of CEMP-no stars (CEMP stars which exhibit no over-abundances of the neutron-capture elements) for four different [Fe/H] bins. Empty squares refer to the fraction of CEMP stars and filled ones to the fraction of CEMP-no stars only, considering a subsample from [10], divided into CEMP-no and CEMP-r/s (CEMP stars which exhibit elemental abundance patterns associated with *rapid* and/or *slow* process) stars following [22] (left panel). The simulated data underestimate the observations, even if the comparison must be restricted only to the CEMP-no fraction, since we do not consider mass-transfer effects in stellar binaries.

[23] have estimated that a fraction of dim SNe as large as 30%-50% is required to reconcile the observed SN rate with that predicted from the observed cosmic SFR. Assuming a similar fraction of Pop II faint-SNe we can better match the observed data points, as shown in Fig. 4.

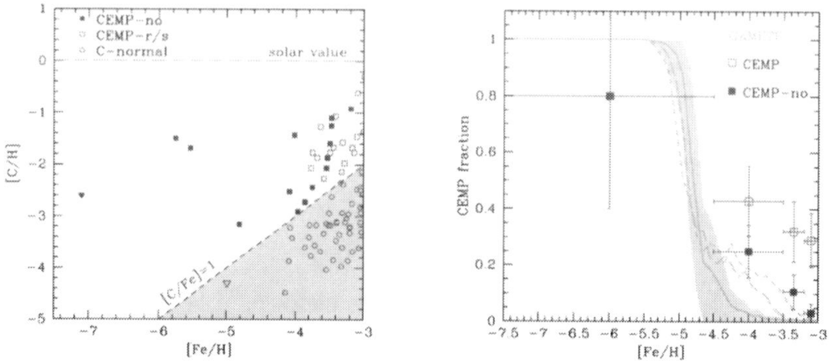


**Figure 4.** *Left panel:* C abundance as a function of [Fe/H] for the sample from [10, 22]. *Right panel:* observed fraction of CEMP (empty squares) and CEMP-no (filled squares) stars from [10, 22]. Errorbars are Poissonian errors (y-axis) and bin amplitude (x-axis). The grey solid line shows the predicted fraction of CEMP-no stars for the fiducial model averaged over 50 merger trees. The shaded region shows the 1-$\sigma$ dispersion among different merger histories. The long-dashed and dashed lines show the results obtained when 30% and 50% of Pop II SN progenitors are assumed to explode as faint SNe, respectively.

## 5. Conclusions

We have presented an implemented version of the semi-analytical code GAMETE and a first applications of this new model to Stellar Archaeology. For the first time, we follow the star formation history and the enrichment in both metals and dust of the MW galaxy. Using a 2-phase description of the ISM we can follow the evolution of the atomic and molecular gas components and the separate evolution of metals and dust grains in these two phases. The model has been calibrated so as to reproduce the observed properties of the MW at $z = 0$ and we find that the simulated progenitors of the MW are in very good agreement with the scaling relations inferred from observations of local and GRB-host galaxies. When applied to Stellar Archaeology, the simulated MDF is very sensitive to the adopted Pop III IMF and metal yields. Current observations seem to require a largely sub-dominant contribution from PISNe to early metal enrichment, favouring Pop III stars with masses in the range $[10-140]M_\odot$ which explode as faint SNe. The analysis on the C abundance confirms the expectations that C-normal stars with $[Fe/H] < -4.5$, such as SDSS J102915, can form only assuming a dust driven transition.

Our fiducial model predicts a steeper decline of the CEMP fraction with [Fe/H] and an over-abundance of C-normal stars in the low-metallicity tail of the MDF compared to observations. This suggests that additional physical processes must be at play, most likely radiative feedback associated to the rising UV-background.

## Acknowledgments

The research leading to these results has received funding from the European Research Council under the European Unions Seventh Framework Programme (FP/2007-2013) / ERC Grant Agreement n. 306476. This research was supported in part by the National Science Foundation under Grant No. NSF PHY11-25915. SS acknowledges support from Netherlands Organization for Scientific Research (NWO), VENI grant 639.041.233

## References

[1] Bromm V, Ferrara A, Coppi P S and Larson R B 2001 *MNRAS* **328** 969
[2] Schneider R, Omukai K, Bianchi S and Valiante R 2012 *MNRAS* **419** 1566
[3] Frebel A and Norris J E 2013 *Planets, Stars and Stellar Systems* vol 5, ed T Oswalt and G Gilmore (Springer) chapter 3
[4] Caffau E, Ludwig H-G, Steffen M, Freytag B and Bonifacio P 2011a *Solar Phys.* **268** 255
[5] Caffau E *et al.* 2011b *Nature* **477** 67
[6] Schneider R, Valiante R, Ventura P, dell Agli F, Di Criscienzo M, Hirashita H and Kemper F. 2014 (arXiv:1404.7132)
[7] Corbelli E, Bianchi S, Cortese L, Giovanardi C, Magrini L, Pappalardo C, Boselli A and Bendo G J 2012 *A&A* **542** A32
[8] Rémy-Ruyer A *et al.* 2014 *A&A* **563** A31
[9] Zafar T and Watson D 2013 *A&A* **560** A26
[10] Yong D *et al.* 2013 *ApJ* **762** 26
[11] Salvadori S, Schneider R and Ferrara A 2007 *MNRAS* **381** 647
[12] Raiteri C M, Villata M and Navarro J F 1996 *A&A* **105** 315
[13] Hosokawa T and Omukai K 2009 *ApJ* **703** 1810
[14] Woosley S E and Weaver T A 1995 *ApJ* **101** 181
[15] Bianchi S and Schneider R 2007 *MNRAS* **378** 973
[16] Marassi S, Schneider R, Limongi M and Chieffi A 2014 *ApJ* **794** 100
[17] Heger A and Woosley S E 2002 *ApJ* **567** 532
[18] Schneider R, Ferrara A and Salvaterra R 2004 *MNRAS* **351** 1379
[19] Van den Hoek L B and Groenewegen M A T 1997 *A&A* **123** 305
[20] Zhukovska S, Gail H P and Trieloff M 2008 *A&A* **479** 453
[21] de Bennassuti M, Schneider R, Valiante R and Salvadori S 2014 *MNRAS* accepted
[22] Norris J E *et al.* 2013 *Apj* **762** 28
[23] Horiuchi S, Beacom J F, Kochanek C S, Prieto J L, Stanek K Z and Thompson T A 2011 *ApJ* **738** 154

Journal of Physics: Conference Series **566** (2014) 012011 doi:10.1088/1742-6596/566/1/012011

# Optical tests of the LUTIN Fabry-Pérot prototype

**L Giovannelli[1], F Berrilli[1], D Del Moro[1], V Greco[2], R Piazzesi[1], A Sordini[2] and M Stangalini[3]**

[1]Dipartimento di Fisica, Università di Roma "Tor Vergata", Via della Ricerca Scientifica, I-00133 Rome, Italy

[2]Istituto Nazionale di Ottica Applicata - CNR, Largo Enrico Fermi 6, I-50125 Firenze, Italy

[3]INAF-Osservatorio Astronomico di Roma, Via Frascati 33, I-00040 Monte Porzio Catone (RM), Italy

E-mail: giovannelli@roma2.infn.it

**Abstract.** The soLar group University of Tor vergata fabry-pérot INterferometer (LUTIN) is a narrow band filter based on an optical cavity resonator with Capacitance-Stabilised Etalon (CSE) control. The prototype, developed at the University of Rome Tor Vergata, is part of the study for the narrow band channel of the ADvanced Astronomy for HELIophysics (ADAHELI) mission designed to investigate the dynamics of solar atmosphere as part of the Italian Space Agency (ASI) Small Missions program. We developed the electro-mechanical control for the optical cavity, necessary for the tuning and the gap control of the instrument. We present the measures of the microroughness of the optical plates, performed with a Zygo interferometer, and the instrument spectral stability behaviour in on-optical-bench open-air mode. The measures refer to the upgraded version of the LUTIN prototype, which mounts the new $\lambda/60$ optical plates.

## 1. Introduction

The Tor Vergata Fabry-Pérot interferometer [1-2] is a narrow-band tunable filter, optimized in the range 500 $nm$ – 800 $nm$, designed and realized in the framework of the phase-A study of the ADvanced Astronomy for HELIophysics (ADAHELI) solar mission project awarded by the Italian Space Agency (ASI) small missions program [3-5].

The chromosphere is the region of the Sun's atmosphere in between the photosphere and the corona. Verification of new models for coronal heating requires the capability of observing both the magnetic fields and plasma motions from the photosphere through the chromosphere at high spatial and temporal resolutions [6-7]. Spectro-polarimetric imagers, such as those based on Fabry-Pérot interferometers, provide the necessary time resolution to disentangle magnetic fields and plasma dynamics [8].

The interferometer is based on an optical cavity (OC) where the incoming light undergoes multiple-beam interference bouncing between the two semi-reflective surfaces of the OC. The spectral output has several peaks of transparency at different wavelengths. The peaks are equispaced, resembling a comb. Parallelism, distance control and stability of the OC are crucial parameters of the interferometer, together with the optical quality. In the realized prototype, they are obtained via a double electro-mechanical positioning system, using two kinematic mounts coupled together and controlled with electronic micrometric screws and piezoelectric actuators. The stability is guaranteed by capacitive sensors measuring the position of the

optical housing and performing a closed loop control on the piezoelectric actuators, making the prototype a Capacitance Stabilized Etalon (CSE)[2].
In this work we present first spectral stability measurements obtained in on-optical-bench open-air mode at Solar Physics laboratory in Tor Vergata University. Results are shown in section 3. Absolute measure of the surface of the optical flat plates has been obtained with a Zygo interferometer at the CNR-Istituto Nazionale di Ottica in Florence, results are reported in section 4.

## 2. The Fabry-Pérot interferometer characteristics

Recently, a couple of fused silica optical plates, $25.4mm$ of diameter, with a nominal surface quality of $\lambda/60$ and a peak-to-valley (PV) surface error of 10 $nm$ has replaced the initial couple of optical flats installed in the interferometer ($\lambda/30$ at 632.8 $nm$ and PV error = 21 $nm$). The plates have been realized and characterized by the manufacturer Lightmachinery.
Coarse adjustment of micrometric screws and piezoelectric position control are performed via LabView programs; the programs transform pitch and yaw numerical inputs to signals to the three 120° symmetry actuators. A further procedure regulates the cavity parallelism with a typical resolution of $\simeq$ 20 $arcsec$. To preserve the two plates from accidental contacts there are software safety parameters and a Teflon ring of 0.5 $mm$ thickness on the boundary of the movable optical plate [9].
Capacitive sensors are used to measure the position of the metal ring retaning the movable optical plate with an error less then 0.5 $nm$. The servo loop provided by the analog electronic controller maintains the movable plate in position using the piezoelectric actuators and the feedback from the capacitive sensors. Piezoelectric actuators are controlled via a digital to analog converter board giving inputs to the analog controller, providing OC gap variations, with a minimum resolution of $\simeq$ 2 $nm$ and a maximum stroke of 15 $\mu m$.

## 3. The Fabry-Pérot interferometer spectral stability

We present the measures of the spectral stability of the interferometer at the Solar Physics Laboratory in the University of Rome Tor Vergata. A sodium lamp extended source is used to illuminate the prototype, with a lens collimating the light coming from the lamp. The collimated beam passes through the prototype and is then focused by a second lens to obtain an image of the fringe system. The sodium lamp provides an incoherent almost monochromatic source of light at $\lambda_0 = 589$ $nm$.

Refraction index of the air inside the cavity varies due to changes in temperature, pressure and humidity. We select 10 hours from a 24 hours dataset, searching for a period with maximum air conditions stability. We obtain from different sensors:

$$T = 22.9 \pm 0.2\ °C$$
$$P = 1016.0 \pm 0.5\ hPa$$
$$RH = 43\ \% \pm 3\%.$$

We use the modified Edlén Equation [10] to compute the refraction index of air obtaining a typical value of $n = 1.000270 \pm 10^{-6}$.

From this calculation we obtain over the 10 hours a maximum drift due to changes in the refraction index:

$$\frac{\Delta\lambda_n}{\lambda} = \frac{\Delta n}{n} = 10^{-6}. \quad (1)$$

The capacitive sensors measure the distance between the sensor itself and a sensing point on the steel ring retaining the optical plate [9]. This measure is based on changes in the capacitance due to a variation in the distance between the two plates. Variations of the air permittivity $\epsilon$ result in a spurious signal $\Delta\lambda_\epsilon$. Several effects can affect the wavelength trasmitted by the

Fabry-Pérot filter. Summing up all the effects coming from the variation of: refractive index $\Delta\lambda_n$, physical distance between the plates due to thermal expansion $\Delta\lambda_T$ and permittivity of air $\Delta\lambda_\epsilon$, we expect variations in the trasmitted wavelength of the order [11]:

$$\frac{\Delta\lambda}{\lambda} = \frac{\Delta\lambda_n}{\lambda} + \frac{\Delta\lambda_T}{\lambda} + \frac{\Delta\lambda_\epsilon}{\lambda} \simeq 10^{-6}. \quad (2)$$

The observed ring system (shown in fig.1) is the result of the multiple beam interference inside the cavity and optical setup of the experiment. During the test, the cavity gap is set to $d = 8.2\ mm$, in order to obatain a sufficent number of interference rings inside the field of view of the camera. We use the following interference order:

$$m = \frac{2n}{\lambda_0} d \cos\theta \simeq \frac{2n}{\lambda_0} d \quad (3)$$

where $\theta$ is the incidence angle and is nearly zero. In our case we have $m = 27851$.

The Free Spectral Range (FSR) is defined as the distance between two peaks in the comb pattern of the spectral response of the instrument [12]. This quantity is equal to:

$$FSR = \frac{\lambda_0^2}{2nd\cos\theta} = \frac{\lambda_0}{m} = 21.15\ pm. \quad (4)$$

During the experiment several images of the fringe system, one every 5 minutes for 10 hours, were obtained. In order to identify the peaks, we compute associated profiles from the center (corresponding to the optical axis of the system) in a chosen radial direction. Then we performe a simple parabolic fit on the first two maxima, as shown in the right panel of fig.1 and we compute the distance between them, expressed as the physical distance in $\mu m$ on the sensor, obtaining the value:

$$D = 2.92\ 10^{-4}\ m. \quad (5)$$

We measure the position of the first ring in every image and, subtracting the value from the previous image, we obtain the variation in the radius $\Delta r$. To compute the spectral shift we convert this value in wavelength shift:

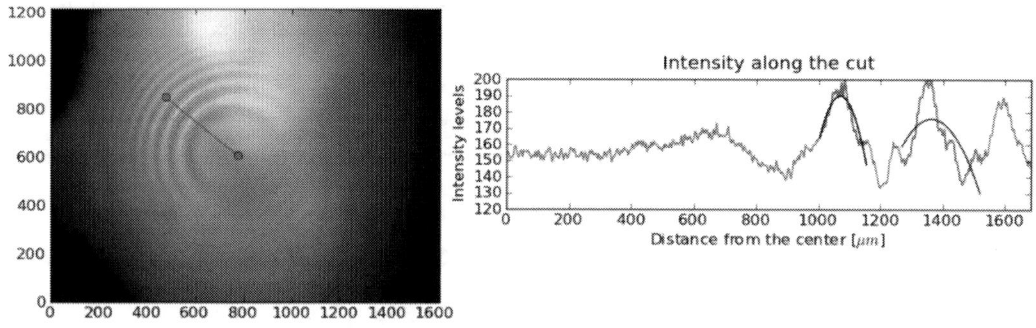


**Figure 1.** Interference Fabry-Pérot pattern. Cut along the red line is shown as an intensity profile in the right panel. We find the radius of the first two maxima rings in every frame using a parabolic fit with a moving window based on previous image (data shown in a different colour). In some images this results in a poor fit of the gaussian width of the second peak, but with a still robust estimation of its position. Nevertheless only the position of the first peak is used for $\Delta r$ calculation.

$$\Delta\lambda = \frac{2n}{m}\frac{\Delta r}{D}\frac{\lambda_0}{2n} = FSR\frac{\Delta r}{D}. \tag{6}$$

We can now obtain the spectral response shift as a function of the radius variations. We plot the relative spectral variations $\frac{\Delta\lambda}{FSR}$ in fig.2. We obtain:

$$\langle\Delta\lambda\rangle = 8.7\ 10^{-14}\ m. \tag{7}$$

The typical RMS over one hour is:

$$RMS = 9.5\ 10^{-13}\ m. \tag{8}$$

We thus obtain a typical $\frac{\Delta\lambda}{\lambda_0} = 1.6\ 10^{-6}$ in agreement with the expected value. The PV relative shift does not exceed 0.24 over one hour. If we multiply by the $FSR$, see eq. 4, the asolute shift in lambda does not exceed a PV error of 5 $pm/h$ over 10 hours.

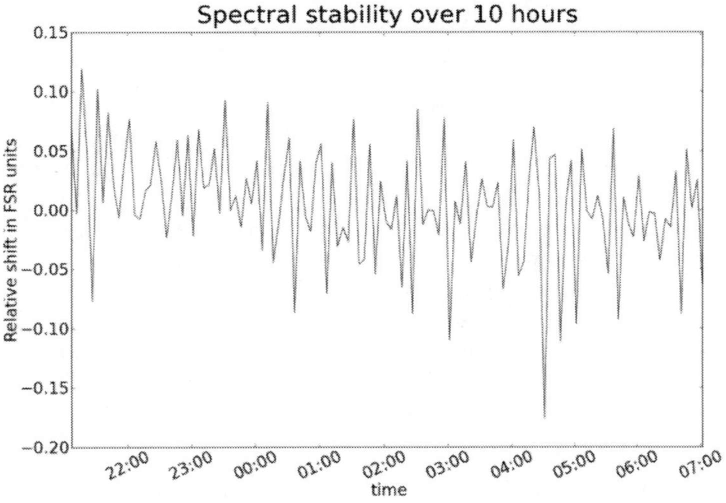


**Figure 2.** Spectral stability over 10 hours. The relative shift in lambda is shown as a function of time. The cadence is 5 minutes.

## 4. The Fabry-Pérot interferometer optical characterization

The two $\lambda/60$ optical plates are fully characterized using two Zygo interferometers at the Istituto Nazionale di Ottica (INO) in Arcetri.

First the two plates have been extracted from the prototype for an absolute measure of their relative surface errors. We use a phase-shift Zygo interferometer operated with a laser source @632.8 $nm$. The absolute shape of the reference plate is measured with the three-planes technique [13] and is subsequently subtracted from the relative measures. We show in fig.3 the surface errors of the two plates. The full diameter is 25.4 $mm$ for both the plates. Measures are referred to a useful diameter of 23 $mm$. For the movable plate, we measure a PV error of 9 $nm$ and an RMS error of 1.603 $nm$, once we subtract the reference. For the fixed plate, we measure a PV error of 8 $nm$ and an RMS error of 1.504 $nm$, once we subtract the reference. The two plates are therefore, respectively, a $\lambda/70$ and a $\lambda/80$ optical quality plates, sligthly better than what declared by the manufacturer.

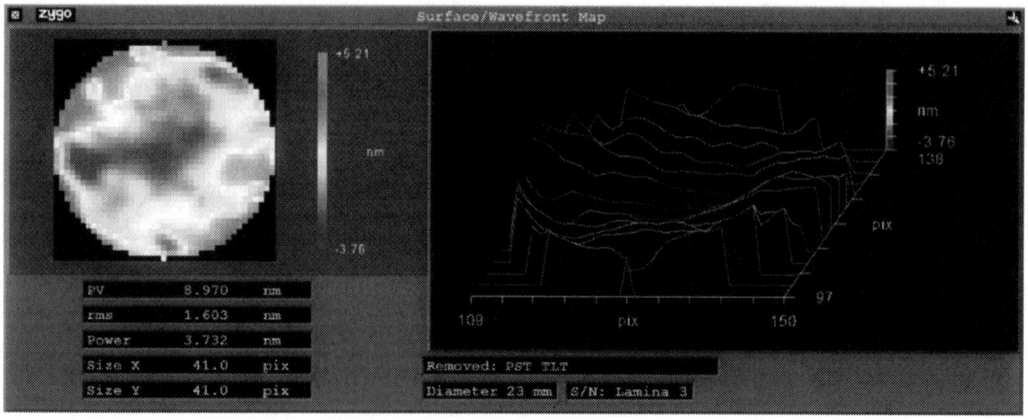


**Figure 3.** Zygo output of the phase-shift analysis. Surface errors and shape measured of the movable optical plate. PV errors refer to the measures obtained without subtracting the absolute shape of the reference plate.

## 5. Conclusions

We obtained the spectral stability characteristics of the LUTIN prototype over a period of 10 hours. We also characterized the two surfaces of the optical cavity, measuring the absolute shape of the two plates.

In order to measure the dynamics of the solar photosphere and taking into account the typical doppler shift due to Earth motion, it is necessary to have an instrument able to discriminate a velocity signal of 50 $m/s$ [14]. This implies a spectral stability of at least 1 $pm/h$. In this open-air configuration, the typical RMS shift for the passband of this prototype @ 589 $nm$ is 0.95 $pm$ over an hour and it does not exceed a PV error of 5 $pm/h$ over 10 hours. Furthermore, it can be inserted in a vessel able to control temperature, humidity and pressure, in order to improve the stability. After the construction of the stabilized vessel, more spectral tests are planned.

## Acknowledgments

This work was supported by ASI Phase A Contract I/020/08/0 ADAHELI.

## References

[1] Berrilli F *et al* 2011 *Society of Photo-Optical Instrumentation Engineers (SPIE) Conference Series* **8148**
[2] Giovannelli L *et al* 2012 *Society of Photo-Optical Instrumentation Engineers (SPIE) Conference Series* **8446**
[3] Berrilli F *et al* 2008 *European Solar Physics Meeting* **12** 6
[4] Berrilli F *et al* 2009 *Mem. SAIt* **80** 251
[5] Greco V Cavallini F Berrilli F 2010 *Society of Photo-Optical Instrumentation Engineers (SPIE) Conference Series* **7731**
[6] Del Moro D *et al* 2007 *Astronomy and Astrophysics* **472** 599
[7] Stangalini M *et al* 2011 *Astronomy and Astrophysics* **534** A65
[8] Berrilli F *et al* 2010 *Advances in Space Research* **45** 1191
[9] Giovannelli L *et al* 2012 *Journal of Physics Conference Series* **383**
[10] Bönsch G and Potulski E 1998 *Metrologia* **35(2)** 133
[11] Hicks T R Ready N K and Atherton P D 1984 *Journal of Physics E Scientific Instruments* **17** 49
[12] Hernandez G 1988 Fabry-Perot interferometers *Cambridge University Press*
[13] Greco V *et al* 1999 *Applied Optics* **38** 2018
[14] Cavallini F *et al* 2000 *The Solar Cycle and Terrestrial Climate, Solar and Space weather, ESA SP* **463** 607

Journal of Physics: Conference Series **566** (2014) 012012 doi:10.1088/1742-6596/566/1/012012

# Shining light on CuO for exploring high-Tc multiferroics

**A. Stucky, A. Ubaldini, J. Levallois, M. K. Tran, D. van der Marel and E. Giannini**

Departement of Quantum Matter Physics, University of Geneva, Switzerland

E-mail: `Adrien.Stucky@unige.ch`

**Abstract.** Searching for a 3D multiferroic material with a strong magnetoelectric coupling and high critical temperature is a major challenge in modern condensed matter research. CuO is the building block of high-temperature superconductors and triggered a new interest when it was established as potential high temperature multiferroic. We have succeeded in growing high quality single crystals of CuO with two different methods, namely the floating zone under high oxygen pressure and the chemical vapor transport growth. The fact that we are able to grow crystals of the same compound by different techniques makes it possible to study the effect of slightly different chemical compositions, various kinds of defects and variable strain on the final properties of the compound. Optical spectroscopy has been deployed to study the optical response of cupric oxide. Thereby we achieved a deeper insight of the optical, electronic and structural properties by measuring the infrared reflectivity under a magnetic field and the Raman shift under hydrostatic high pressure.

## 1. Introduction

According to its definition, a multiferroic material is a solid phase where two or more primary ferroic properties are united. In our case we have focused on magnetoelectric materials where an intrinsic magnetization can be induced by applying an electric field or inversely, where an electric polarization can be created by applying an external magnetic field. For being exploitable multiferroicity must occur at high temperature and CuO appears to be one of the few candidates to show multiferroics properties close to room temperature [1].

Cupric oxide, CuO, is one of the two stable copper oxides, the other being $Cu_2O$. Even if its chemical formula is very simple, this material is highly interesting because of its structural, chemical and physical properties. CuO is the only monoxide transition metal who crystalizes in a monoclinic crystal system [2], belonging to space group $C2/c$ [3]. The structure of CuO can be viewed as being composed of two types of zigzag $Cu - O$ chains running along two different crystallographic directions. Cupric oxide recovered a new surge of interest since it was established as a joint-order parameter multiferroic compound, where a non-collinear spiral magnetic order breaks inversion symmetry, thus inducing ferroelectricity. In most cases, this spiral-phase-induced ferroelectricity exists only at low temperature, but it was found that for CuO, magnetic transitions occur at 213 $K$, 229.3 $K$ and 230 $K$ [4].

## 2. Crystal growth

It is well accepted that the credibility of a result in condensed matter experiments is strongly correlated to the quality of the crystal. We successfully grew our own bulk crystals using two different techniques: the optical floating zone method ($OFZ$) in an image furnace and the chemical vapor transport ($CVT$) process in sealed quartz tubes. Thanks to both techniques we obtained high quality macroscopic crystals with facets larger than several squared millimeters.

### *2.1. Optical floating zone method*

In the first case, commercial powders of high purity (>99.99%) CuO were hydrostatically pressed into a cylindrical rod under 40 $MPa$. In order to increase the density of the precursors rod we established a grain size analysis. We found that the density of the rods was a major parameter for finding the more stable conditions during the growth. We sintered the rod at 900°$C$ for 72 hours then the rod was cut into two parts: one being about 6 $cm$ long and the second about 2 $cm$. The long one was hung on the upper shaft and the short one was held on the lower shaft (in order to be used as a seed) of an infrared radiation furnace equipped with two ellipsoidal mirrors. We used 1000 $W$ halogen lamps as light sources. The rods are inserted inside a quartz chamber filled with 9 $bars$ of $O_2$ at the common focus of both mirrors. By heating the rod ends with the light emitted by the lamps, a molten zone formed that connected the rods to each other. The strong thermal gradient induced a crystallization front in the lower liquid-solid interface when the two parts travel downwards at a controlled velocity. The homogeneity of the liquid was increased by counter rotating the upper rod at 25 $rpm$ and the lower rod at 20 $rpm$. A crystallized rod was grown using a slow rate of 5 $mm/h$. Under these conditions, a crystal rod of 4 $cm$ length and 5 $mm$ in diameter was grown.

### *2.2. Chemical vapour transport process*

Using the chemical vapor transport method, a mixture of high purity metallic copper was mixed in stoichiometric ratio with $BaO_2$ according to the formal reaction: $Cu + BaO_2 \rightarrow CuO + BaO$. The transport agent $I_2$ was added to the mixture, with a ratio $I_2$:Cu ranging between 0.01 and 0.4. These reagents were sealed in a quartz ampoule, 12 $cm$ long and 0.8 $cm$ inner diameter, under high vacuum equal to $10^{-5}$ $mbar$. The sealed ampoule was placed in a resistive furnace with a specific thermal gradient of approximately 6.5 $K/cm$. The extremity containing the reagents was heated at 950°$C$ and the other side at 820°$C$ during 10 days. The system was then slowly cooled down to room temperature. After breaking the quartz ampoule, the excess of iodine was washed out by several acetone or ethanol bath and some high purity CuO macroscopic bulk crystals were collected in the cold extremity.

### *2.3. Characterization*

The single crystals were characterized by X-Ray diffraction whose results are presented in Fig. 1.

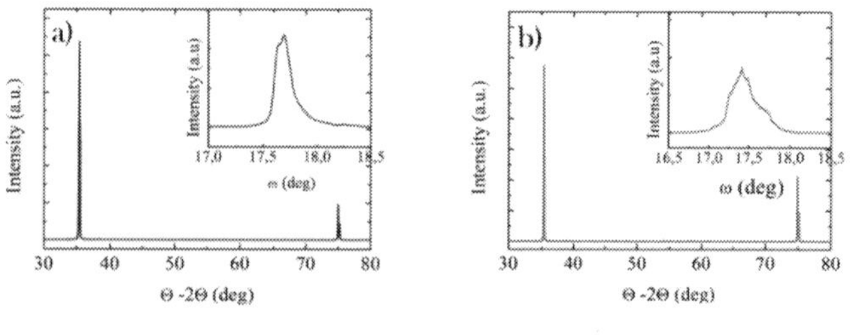


**Figure 1.** a) XRD pattern on single crystal grown by OFZ. The inset is the rocking curve. b) XRD pattern on single crystal grown by CVT.

We recognize one single orientation in both spectra. By reproducing the measurement on several crystal, we established that both techniques lead to crystals with high quality of crystallization and a preferred cleavable orientation in [ 111] direction.

Journal of Physics: Conference Series **566** (2014) 012012 doi:10.1088/1742-6596/566/1/012012

## 3. Optical spectroscopy

We successfully used optical spectroscopy to study the natural frequencies of cupric oxide. In this work, we achieved a deeper understanding of its optical, electronic and structural properties, by measuring the infrared reflectivity under a magnetic field and the Raman shift under hydrostatic high pressure. A large frequency range was under our control to derive the optical modes and define their behavior at the occurrence of magnetic transitions.

### *3.1. Infrared spectroscopy*

We performed reflectivity measurements on a CuO sample, grown by $OFZ$ and cleaved in the $[-111]$, at various temperatures from 200 $K$ until room temperature and external magnetic field up to 7 $T$ in the far infrared range from 50 to 650 $cm^{-1}$. The sample was mounted on a coldfinger placed in a superconducting magnet which is attached to a Fourier Transform Infrared Spectrometer. The absolute reflectivity, presented in the Fig. 2, was obtained by calculating the ratio between the intensity of the reflected beam on the sample and the reflected intensity of a mirror. Polarized incoming light allows us to extract the reflectivity of different planes in the crystal by interpolating the original data as explained in [5]. Data of reflectivity in the range 50 to 650 $cm^{-1}$, attributed to two different planes, are showed in Fig. 2 a) and b) and are in good agreement with the literature [6].

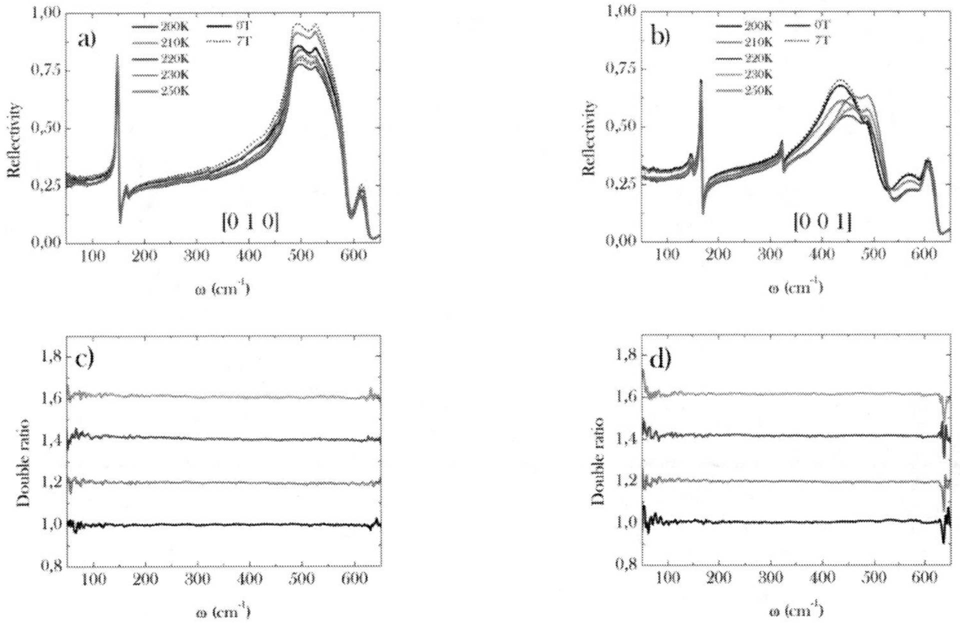


**Figure 2.** Reflectivity of one CuO single crystal, extracted for the [010] (panel a) and [001] (panel b) crystallographic planes. Solid lines show spectrum measured at zero magnetic field. Dotted lines show spectra measured at 7T. In panels (c) and (d), the intensity ratio between reflectivity at zero field and at 7T (called *Double ratio)* is shown, for the [010] and [001] plane, respectively.

By plotting the ratio between the intensity measured at 7 $T$ and at zero field for different temperatures, we can probe the effect of the magnetic field on the infrared active optical modes. Result of this analysis is shown in Fig. 2 c) and d). For sake of visibility curves are shifted on vertical axis. We observed that the active infrared modes are not sensitive to an external mangetic field up to 7 $T$.

### 3.2. Raman scattering
We performed Raman spectroscopy on single crystals of CuO grown by $OFZ$. The temperature was ramped from 150 $K$ to room temperature and the evolution of the Raman response with T is shown in Fig. 3. The three expected Raman active modes are clearly visible and show very high sharpness and proper intensity according to [7]. We notice the appearance of two small peaks between 200 and 250 $cm^{-1}$ in the Raman shift below the temperature of the magnetic transition. These two peaks confirm the previous observations by Chen *et al.* [8] and are attributed to zone-folded phonons.

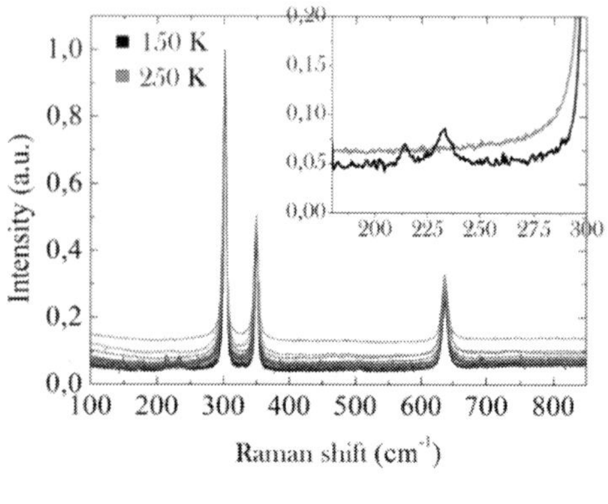


**Figure 3.** Raman shift spectrum of CuO as a function of temperature from 150 to 250$K$. Zoom at low frequency in the inset.

A study in high hydrostatic pressure conditions at room temperature was performed on microscopic single crystals grown by $OFZ$. We used a Diamond Anvil Cell filled with Daphnee oil 7373 to reach 25 $GPa$ as described elsewhere [9]. The results of this study are shown in Fig. 4. The behavior of the optical modes under pressure is in very good agreement with the literature[10].

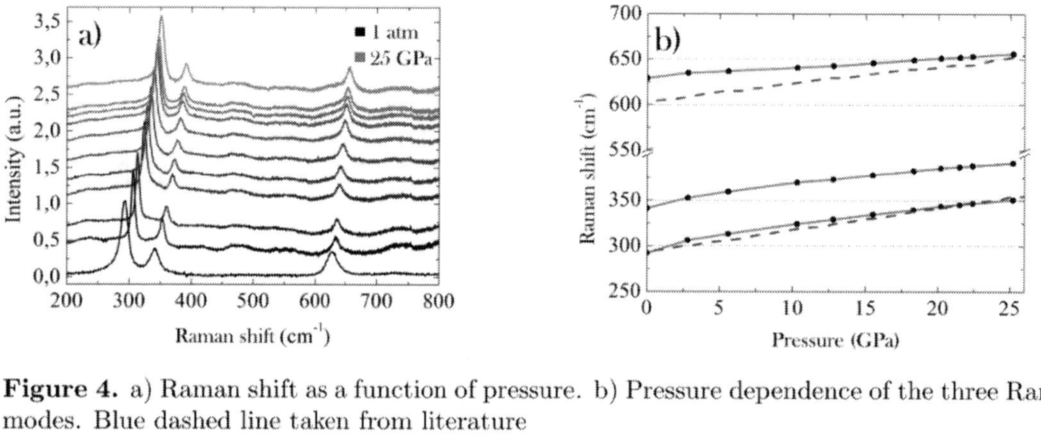


**Figure 4.** a) Raman shift as a function of pressure. b) Pressure dependence of the three Raman modes. Blue dashed line taken from literature

## 4. Conclusion
We have grown high quality single crystals with two different techniques, the optical floating-zone and the chemical vapor transport. A high purity and high crystallinity have been obtained

for samples grown by $OFZ$. We have carried a study of IR and Raman spectroscopy under pressure, magnetic field and temperature of the crystal face $[-111]$ for different polarizations. We confirm the appearance of extra modes below $T_N$ as expected for the monoclinic structure of CuO. We confirm that the $C2/c$ phase is stable up the 25 $GPa$. We also observe that magnetic fields up to 7 $T$ for $T = 200$ $K$ do not change the multiferroic phase. In particular the phonons shift much less than 5 $cm^{-1}$, if at all.

## References

[1] Kimura T, Sekio Y, Nakamura H, Siegrist T and Ramirez A P 2008 *Nature Materials* **7** 291–294
[2] Åsbrink S and Norrby L J 1970 *Acta Crystallographica Section B: Structural Crystallography and Crystal Chemistry* **26** 8–15
[3] Forsyth J B, Brown P J and Wanklyn B M 1988 *Journal of Physics C: Solid State Physics* **21** 2917–2929
[4] Villarreal R, Quirion G, Plumer M L, Poirier M, Usui T and Kimura T 2012 *Physical Review Letters* **109** 167206
[5] Kuzmenko A B, Tishchenko E A and Orlov V G 1996 *Journal of Physics: Condensed Matter* **8** 6199–6212
[6] Kuzmenko A, van der Marel D, van Bentum P, Tishchenko E, Presura C and Bush A 2001 *Phys. Rev. B* **63** 094303
[7] Hagemann H, Bill H, Sadowski W, Walker E and François M 1990 *Solid State Communications* **73** 447–451
[8] Chen X K, Irwin J C and Franck J P 1995 *Phys. Rev. B* **52**(18) R13130–R13133
[9] Tran M K, Levallois J, Lerch P, Teyssier J, Kuzmenko A B, Autès G, Yazyev O V, Ubaldini A, Giannini E, van der Marel D and Akrap A 2014 *Physical Review Letters* **112** 047402
[10] Wang Z, Pischedda V, Saxena S K and Lazor P 2002 *Solid State Communications* **121** 275–279

Journal of Physics: Conference Series **566** (2014) 012013 doi:10.1088/1742-6596/566/1/012013

# Size dependence of excited state properties in zigzag graphene nanoribbons

**Jessica Alfonsi**

The Abdus Salam International Centre for Theoretical Physics (ICTP), Strada Costiera 11, I-34151 Trieste, Italy

E-mail: jalfons1@ictp.it, jessica.alfonsi@gmail.com

**Abstract.** Density-matrix renormalization group results are reported for the excitonic transitions in zigzag graphene nanoribbons within the framework of the simple Hubbard model. This work shows that a one-dimensional quantum lattice model with open boundary conditions is able to capture the many-body features in the low-energy electronic spectra of these systems alongside the size dependence on the ribbon width. The obtained results agree with the trends of charge and spin excitation gaps available from quantum chemistry predictions.

## 1. Introduction

The latest decade has seen a dramatically increasing interest into two-dimensional (2D) graphene for its unique properties which make it a promising material for future electronics [1]. However, one of the main limits preventing the application of 2D monolayer graphene in devices such as field-effect transistors is given by the absence of an OFF state due to its gapless electronic structure. Recently, quasi-one-dimensional (1D) stripes of graphene, also known as graphene nanoribbons (GNRs), have attracted significant interest for the possibility of opening energy gaps in the electronic structure of graphene according to the width and the edge geometry of the chosen ribbon structure [2]. Lately, new strategies have been devised for the synthesis of GNRs by attempting to control their size and edge chirality, ranging from the pioneeristic top-down lithographic methods to bottom-up techniques exploiting aromatic precursors as well as chemical and laser unzipping of carbon nanotubes (CNTs) [3].

Beside armchair-edged graphene nanoribbons (AGNRs), widely investigated for their family-behaviour in energy gaps [4], zigzag graphene nanoribbons (ZGNRs) have attracted the focus of researchers for their spin-related electronic structure, which shows a spatial separation on opposite edges of electrons at the Fermi level. This suggests a spin configuration of these electrons with ferromagnetic coupling along each edge but with antiferromagnetic one between opposite sides [5], which would lead to interesting optical properties [6]. The quasi-1D geometry of GNRs enhances Coulomb interactions between electrons and holes, giving rise to excitonic states, which can deeply affect the optical properties of these nanostructures and consequently the design of new devices.

In this work the half-filled Hubbard model Hamiltonian mapped to an open-ended 1D quantum lattice model is used to provide a full many-body treatment of the excited-state properties of ZGNRs going beyond mean-field methods [5] and many-body perturbation theory [6]. Simulation results based on density-matrix renormalization group (DMRG) algorithm [7, 13] will be shown in order to address the low-energy excitonic fine structure and the size dependence of related optical excitations in these systems.

## 2. Quantum lattice models for graphene nanoribbons

Tight-binding (TB) method has been widely used to understand the basic aspects of electronic structure in graphene-based systems such as graphene nanoribbons both within reciprocal [8] and real space frameworks [9]. In the case of AGNRs, TB method has been able to elucidate the size-dependent family behaviour of their electronic band structure. Such findings were obtained by considering the topological equivalence of AGNRs to brick-type lattice models where the hexagonal symmetry of the honeycomb lattice has been neglected, and in turn to two-leg ladder models with uniform TB hopping parameter [9]. This model was successfully employed for addressing the size-dependence of $k = 0$ excitonic states in these systems within the framework of the simple Hubbard model [4].

Fig.1 shows the structure of a zigzag GNR with the primitive unit cell highlighted on the honeycomb lattice. The system is composed of $N$ carbon lines joining A- and B-type carbon atoms, thus comprising a total number of $2N$ sites. From Fig.1 it appears that each edge is composed of atoms entirely of the same type, i.e. A (B) type on the upper (lower) edge, unlike AGNRs for which both types are exposed on each side of the ribbon.

It was shown that the $2N \times 2N$ $k$-dependent Hamiltonian can be mapped to that of a $2N$-site open chain with alternating hopping parameters between atomic-like wavefunctions: $t$ and $2tc\,(k)$, where $c\,(k) = \cos\,(ka/2)$. By numerical diagonalization of the related hamiltonian matrix one can obtain energy bands with peculiar features, such as two partially flat degenerate zero-energy bands between the Dirac points ($k = \pm 2\pi/3a$) and the edge of the Brillouin zone (BZ) with the corresponding states mainly localized at the ribbon edges. Moreover, $c\,(k) = 0$ at the BZ edge, so the bands are highly degenerate, whereas at the Dirac point $c\,(k = \pm 2\pi/3a) = 1/2$, so the quantum lattice model to be considered becomes equivalent to an open-ended chain with uniform hopping $t$ everywhere [9]. This scheme is particularly convenient for performing DMRG simulations applied to the Hubbard model for investigating correlation effects in ZGNRs, since this algorithm best converges for one dimensional systems with open boundary conditions [7].

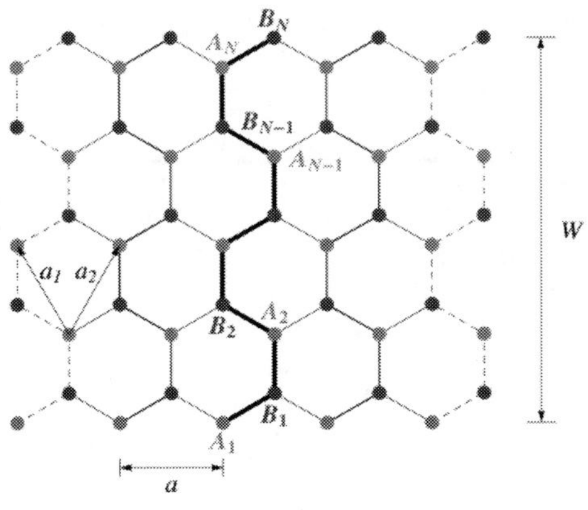


**Figure 1.** Structure of a zigzag graphene nanoribbon with $N$ carbon lines (thick lines) across its width $W$ joining pairs of $A$- and $B$-type carbon atoms. The ribbon is infinite along the horizontal direction (dashed lines). The graphene lattice unit vectors are denoted by $a_{1,2}$ and the lattice parameter by $a$.

## 3. Method

In order to compute the low-energy excited state properties of GNRs the simple Hubbard model for $\pi$ electrons has been considered

$$H = -t_\pi \sum_{\langle i,j\rangle,\sigma} \left(c^\dagger_{i,\sigma} c_{j,\sigma} + h.c.\right) + U \sum_i n_{i,\uparrow} n_{i,\downarrow} \tag{1}$$

where $c^\dagger_{i,\sigma}$ and $c_{i,\sigma}$ are electron creation and annihilation operators on first nearest neighbor sites, $n_{i,\sigma} = c^\dagger_{i,\sigma} c_{i,\sigma}$ is the number of electrons on site $i$ with spin $\sigma$, $t_\pi$ and $U$ are the nearest-neighbour hopping parameter and the on-site Coulomb repulsion parameter between two electrons with opposite spins, respectively. The parameter $U$ must be considered an effective parameter taking implicitly into account also longer range Coulomb interactions in the limit of static screening [4]. Since a low-intermediate regime for the Hubbard correlation coupling strength ($1 \leq U/t \leq 2.2$) seems plausible for graphene-based systems [10, 11], the DMRG calculations reported here were generally performed for the frequently used setting $U/t = 1$. The many-particle basis set was defined for half-filled systems with $n$ electrons distributed over $n = 2N$ sites and total spin quantum number $S_z = 0$. Therefore, the size of the basis set and the dimension of the matrix to be diagonalized is $D = [n!/(n_\uparrow ! n_\downarrow !)]^2$, where $n_\uparrow$ and $n_\downarrow$ are the numbers of spin-up and spin-down electrons, respectively, with $n_\uparrow = n_\downarrow = N$. Since $D$ is already huge for systems with 16 sites (e.g. 8-ZGNR), a numerical solution of the Hubbard Hamiltonian based on Exact Diagonalization (ED) method would be a computationally prohibitive task. Thus DMRG simulations offer a viable alternative to accomplish the investigation of wider ZGNRs, which can be mimicked by 1D systems with open boundary conditions [7]. Moreover, DMRG allows to address correctly quantum fluctuations beyond mean-field approximation, whose validity has been recent matter of debate in graphene literature [11, 12].

The ground state $|\Psi_{GS}\rangle$ of the Hubbard Hamiltonian mapped to an open chain has been obtained through a variational optimization search algorithm based on matrix product states (MPS), which is a class of 1D tensor network states underlying the DMRG method [13], as implemented in the Matrix Product Toolkit code [14]. An iterative eigensolver based on the Lanczos algorithm was used. The obtained results for the smallest 4-ZGNR were validated against ED calculations performed with the ALPS libraries [15].

The low-lying excited eigenstates within each $S_z$ sector were found by a sequential optimization procedure and repeated orthogonalization with respect to the ground state and (whether available) the already obtained excited states lying above it [13]: first the ground state $|\Psi_{GS}\rangle$ is obtained, then the lowest excited state $|\Psi_1\rangle$ that is orthogonal to the ground state i.e. $\langle\Psi_1|\Psi_{GS}\rangle = 0$; after that one looks for a state $|\Psi_2\rangle$ such that both orthogonalization constraints with respect to $|\Psi_1\rangle$ and $|\Psi_{GS}\rangle$ are obeyed, and so on. This differs from traditional DMRG approaches implementing the calculation of excited states through multi-state targetting.

The MPS context also allows to calculate efficiently overlaps and expectation values of the optical conductivity operator, hence optical matrix elements between the ground state and the obtained low-lying excited states $|\langle\Psi_n|v_{\parallel,\perp}|\Psi_0\rangle|^2$, where $v_{\parallel,\perp}$ is the velocity (fermionic current) operator for light polarization either parallel or perpendicular to the ZGNR edges [4].

Other quantities relevant to transport properties which can be easily addressed by DMRG-MPS are the charge gap $\Delta_C$ and spin gap $\Delta_S$ given respectively by

$$\Delta_C = \left[E_0\left(N_e+1, S_0\right) + E_0\left(N_e-1, S_0\right) - 2E_0\left(N_e, S_0\right)\right] \tag{2}$$

$$\Delta_S = \left[E_0\left(N_e, S_0+1\right) - E_0\left(N_e, S_0\right)\right] \tag{3}$$

where $L$ is the number of sites in the chosen lattice model, $N_e$ the number of electrons and $S_0$ the lowest $S_z^{tot}$ component of the total spin such that $S_z = 0$ ($S_z = 1/2$) for even (odd) $N_e$.

Journal of Physics: Conference Series **566** (2014) 012013 doi:10.1088/1742-6596/566/1/012013

## 4. Results

DMRG-MPS simulations were performed for ZGNRs with the number of carbon lines $4 \leq N \leq 20$, i.e. for open chains with the number of sites ranging from 8 to 40.

In Fig.2 the trends of one-photon active excitonic transition energies (left) and related intensities (right) are reported against the ZGNR width expressed by the number of sites in the open chain model. The trend of the transition energies show clearly an inverse dependence on the ZGNR width with only the lowest one-photon active transition for light polarized perpendicular to the ribbon edge $|ES11_{\perp}\rangle$ displaying a non-negligible oscillator strength. However, the magnitude of the oscillator strength of this exciton is lower than those found in AGNRs for both polarization directions. Whereas in AGNRs two optical transitions were found for parallel light polarization and one for light polarized perpendicular to the ribbon edge, in this case the situation is reversed. This result agrees with the well known $\pi/2$ phase shift of the polarization dependence of interband optical transition in ZGNRs with respect to AGNRs (see Refs. 22, 23 in Ref.[4]).

Fig.3 shows the excited state spectrum at the Dirac point of 8-ZGNR obtained with parameters $U/t = 1$ and $t = 2.6\,\text{eV}$, displaying an alternating sequence of singlet and triplet states at low energy. Only one triplet dark (optically inactive) exciton is found here just above the ground state $|GS\rangle$, unlike AGNRs and CNTs, which show a few high-spin states above it.

Remarkably, the two-photon active state denoted as $|TPA2\rangle$ is found to have an oscillator strength value comparable or even slightly higher than the one related to the lowest bright one-photon active exciton $|ES11_{\perp}\rangle$. Such result was not found in AGNRs, where the two-photon intensities were always found to be always lower than the oscillator strength of one-photon active excitons. This finding can be relevant for devising optical limiting applications based on the interesting non-linear optical properties of graphene-based systems [2].

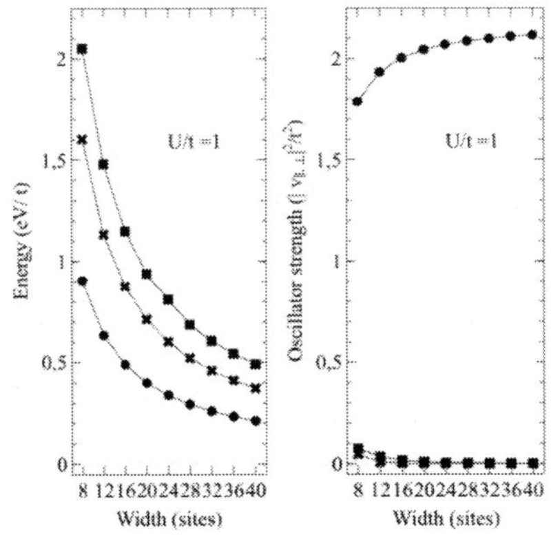


**Figure 2.** Transition energy (left) and optical oscillator strength (right) dependence on ribbon width of one-photon active excitonic states, for light polarization parallel (**x**) and perpendicular (•,■) to the ribbon edge.

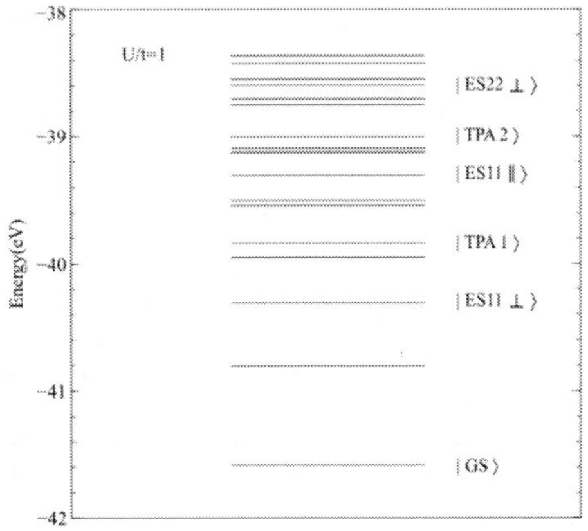


**Figure 3.** Dirac-point excitonic spectrum of 8-ZGNR ($t = 2.6\,\text{eV}$) with one-photon (ES) and two-photon active (TPA) excited states and spin multiplicity discrimination of dark states: singlet (red), triplet (blue), quintuplet (green).

Fig.4 shows again an inverse dependence of charge $\Delta_C$ and spin $\Delta_S$ gaps calculated with Eqq.2,3 (normalized on $t$) on ZGNR width. The obtained values are in agreement with those calculated at the Dirac point by Dutta and Wakabayashi using many-body configuration interaction

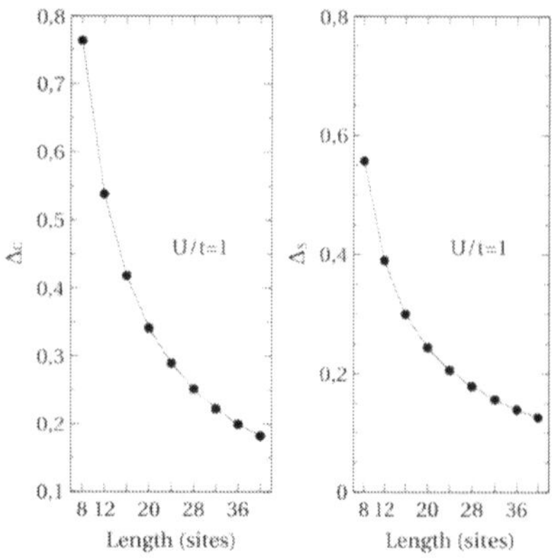


**Figure 4.** Charge gap $\Delta_C$ and spin gap $\Delta_S$ dependence on ribbon width calculated at the Dirac point.

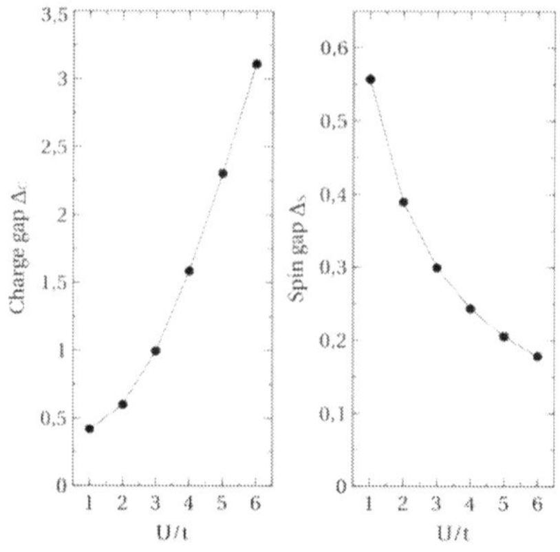


**Figure 5.** Charge gap $\Delta_C$ and spin gap $\Delta_S$ dependence on correlation strength $U/t$ for 8-ZGNR.

with the complete active space approximation (CAS-CI)[10]. Moreover it was verified that $E_0\,(N_e+1, S_0) - E_0\,(N_e-1, S_0) = U$. Fig.5 also confirms their prediction related to the opposite trends of $\Delta_C$ and $\Delta_S$ with increasing Hubbard correlation parameter $U/t$, which supports the possibility of developing spin filtering applications based on the combination of both insulating and magnetic behaviour of ZGNRs by tuning electron correlations.

## 5. Conclusions

In summary, simulation results were reported for the excited state spectra of zigzag graphene nanoribbons based on density-matrix renormalization group algorithm applied to the half-filled Hubbard model within the recently formulated matrix product state formalism. It was shown that the mapping to an open-ended 1D quantum lattice model captures correctly the polarization-dependent optical features of these systems as well as the size- and correlation-dependent trends of charge and spin gaps in agreement with recent results obtained from quantum chemistry configuration interaction methods.
Further analysis of the variationally optimized many-body states is required to inquire the spatial profile of excitons and spin polarization, especially at the ribbon edges, with the help of appropriate correlation functions, which will be the subject of future work.

## Acknowledgments

The author wishes to thank Prof. E. Ercolessi and Dr. C. Degli Esposti Boschi from University of Bologna, Dept. of Physics, for financially supporting her stay in Trieste during the conference.

## References

[1] Geim A K and Novoselov K S 2007 *Nat. Mater.* **6** 183
[2] Kusmartsev F V, Wu W M, Pierpoint M P and Yung K C 2014 Application of Graphene within Optoelectronic Devices and Transistors *Preprint* cond-mat.mtrl-sci/1406.0809v1
[3] Yazyev O V 2013 *Acc. Chem. Res.* **46** 2319
[4] Alfonsi J and Meneghetti M 2012 *New J. Phys.* **14** 053047
[5] Nakada K, Fujita M, Dresselhaus G and Dresselhaus M S 1996 *Phys. Rev. B* **54** 17954
[6] Yang L, Cohen M L and Louie S G 2007 *Nano Lett.* **7** 3112

[7] White S R 1992 *Phys. Rev. Lett.* **69** 2863; Schollwöck U 2005, *Rev. Mod. Phys.* **77** 259
[8] Wakabayashi K, Sasaki K I, Nakanishi T and Enoki T 2010 *Sci. Technol. Adv. Mater.* **11** 054504
[9] Cresti A, Nemec N, Biel B, Niebler G, Triozon F, Cuniberti G and Roche S 2008 *Nano Res.* **1** 361
[10] Dutta S and Wakabayashi K 2012 *Sci. Rep.* **2** 519; *idem* 2014 *Jpn. J. Appl. Phys.* **53** 06JD01
[11] Yazyev O V 2010 *Rep. Prog. Phys.* **73** 056501
[12] Golor M, Wessel S and Schmidt M J 2014 *Phys. Rev. Lett.* **112** 046601
[13] McCulloch I P 2007 *J. Stat. Mech.: Theor. Exp.* **10** P10014; Schollwöck U 2011 *Ann. Phys.* **326** 96
[14] Matrix Product Toolkit `http://physics.uq.edu.au/people/ianmcc/mptoolkit/index.php`
[15] Bauer B et al. (ALPS Collaboration) 2011 *J. Stat. Mech.: Theor. Exp.* 2011 (5) P05001

Journal of Physics: Conference Series **566** (2014) 012014 doi:10.1088/1742-6596/566/1/012014

# Tuning of band gap due to fluorination of graphyne and graphdiyne

**B Bhattacharya, N B Singh, and U Sarkar***

Department of Physics,
Assam University, Silchar-788011, INDIA

E-mail: utpalchemiitkgp@yahoo.com (U. Sarkar).

**Abstract.** The electronic properties of graphyne and graphdiyne consisting of *sp* and $sp^2$ hybridized carbon atom have been investigated within the density functional theory (DFT) method. The corresponding changes in the electronic properties due to systematic functionalization by fluorine at different possible sites are reported. Our band structure calculations clearly infer that all fluorographyne are wide band gap semiconductor and the band gap can be tuned by fluorination and the possibility of modulating the band gap provides flexibility for its use in nanoelectronic devices. Projected density of state (PDOS) analysis provides the clear idea about the bonding nature of these novel materials in details and Crystal Orbital Hamilton Population (-COHP) analysis shed insight on the orbital participating in bonding and antibonding.

## 1. Introduction

Graphene [1] has received a great deal of attention due to its potential application in nanoelectronics. Its importance is attributed to its high mechanical strength, unique electronic properties like high electron mobility and having Dirac cone structure [2, 3]. Opening of band-gap is essential for the successful application of graphene in nano-electronics [4]. Experiments have verified that hydrogenation and fluorination of graphene leads to opening of the band gap [5, 6]. Unfortunately, the H and F atoms on graphene show tendency to aggregate with each other, which makes tuning the band gap extremely difficult [7]. Graphyne, a new layered carbon allotrope consisting of hexagonal rings joined together by acetylenic linkages is also been reported to have Dirac cones [8]. In comparison to graphene, the presence of acetylenic linkages in graphyne provides more flexibility for modulating the band gap. Graphdiyne, another allotrope of carbon, can be considered as an expanded form of graphyne having two acetylenic linkages. Considerable effort has been devoted to the experimental synthesis of graphyne [9] and graphdiyne [10].

In one of our previous studies [11], we have studied the pristine and BN-doped graphyne and other members of graphyne family and proposed a systematic way for widening their band gap. In this study, we investigate the possibility of widening and also tuning the band gap of graphyne and graphdiyne by fluorination based on first-principle calculations.

## 2. Computational details

We performed our first-principle calculations using the density functional theory (DFT) method [12] as implemented in the siesta package [13]. The GGA in Perdew-Burke-Ernzerhof (PBE) scheme are chosen to describe the exchange correlation part of density functional. The DZP basis set has been employed and the norm conserving Troullier-Martins pseudopotentials [14] are used to describe the interaction between cores and valence electron. Relaxation was done until the magnitudes of the forces acting on all atoms are less or equal to 0.040 eV/Å. Brillouin zone was sampled by 11×11×1 k-points with cut off 300 Rydberg. The vacuum space of 15 Å has been used along the z direction in order to avoid the interaction between two images. To study the stability of these systems, the cohesive energy is calculated as

$$\textbf{Cohesive Energy} = \frac{\mathbf{E(System) - [nE(C) + mE(F)]}}{\mathbf{n + m}}$$

where E (System), E (C) and E (F) are energies of the whole system, carbon and fluorine atoms, and n, m are the number of carbon and fluorine atoms present in the system respectively.

## 3. Results and discussions

The optimized geometries are presented in figure 1. The figure 1 (c-e) represents the graphyne with F atoms at linear chain, at ring and at both chain and ring respectively. The F atoms are arranged themselves in alternative carbon atom in such a manner that the structure attains chair like conformation.

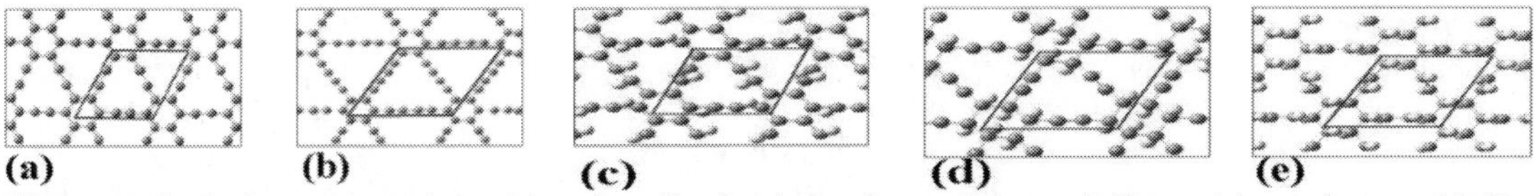

**Figure 1.**Optimize geometries with unit cell of: (a) graphyne; (b) graphdiyne; (c) graphyne with F at chain; (d) graphyne with F at ring; (e) graphyne with F at both chain and ring.

Lattice length of all the considered systems are presented in table S1. It is observed that cell length increases when F atoms are introduced at different site and attains its maximum value when F atoms are introduced in the hexagonal ring site. The possible reason for the observed behaviour is the formation of different types of bonds at different sites.

### *3.1 Band structure*

The calculated band-structure of pristine graphyne with other fluorographynes is plotted in the figure 2 and the band-structure of graphdiyne with other fluorographdiyne is given in supplementary material figure S1. The band gaps of the graphyne and graphdiyne along with their fluorine derivatives are tabulated in table S2 and figure S2 (a) [see in supplementary material].

From the table S2, it is clear that all the flourographynes are wide band gap semiconductor whose band gaps are found to be in the range of 1.647 to 3.750 eV and band gap depend on the functionalization site. We observed that the valence-band maximum (VBM) and conduction-band minimum (CBM) of graphyne (presented in figure 2a) are located at the same (M) point in the hexagonal Brillouin zone for graphyne and at gamma point for graphdiyne, which is a common feature of direct band gap semiconductor and consistent with our previous result [11].

Next, we calculate the band-structure by introducing fluorine atoms at the linear chain positions, at the hexagonal positions and at linear chain as well as in hexagonal positions and are presented in figure 2b, 2c and 2d respectively. The band-structure of graphyne with F atoms at linear chain site (figure 2b) indicates that both VBM and CBM remain at the M point and the band gap is 1.647 eV, which is much greater than that of graphyne. The presence of F atoms at the hexagons of graphyne increased the band gap to 3.750 eV, and the VBM and CBM are now shifted at K point, as shown in figure 2c. However the band-structure for graphyne with F atoms at chain as well as ring given in figure 2d, reveals that VBM (K-point) and CBM (Γ point) are now located at different points of the Brillouin zone with a wide band gap of 3.318 eV, that clearly shows its indirect band gap nature.

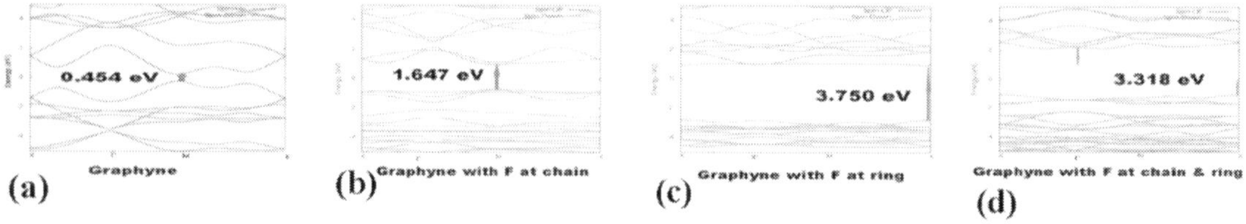


**Figure 2.** Bandstructure of structures: (a)graphyne, (b) graphyne with fluorine at chain (c) graphyne with fluorine at ring (d) graphyne with fluorine at both chain and ring.

We see that graphdiyne has a higher band gap compare to graphyne. Again, when we introduce F atoms at chain or ring site of graphdiyne, the system remains direct band gap semiconductor with VBM and CBM located at Γ point of the Brillouin zone. But whole fluorination of graphdiyne makes the structure an indirect band gap semiconductor.

### *3.2 Cohesive energy*

The more negative cohesive energy signifies the greater stability of the system; it is clear from figure S2b that the pristine systems are more stable than other fluorine functionalized system. For graphyne, the stability on placing F atoms are in the order of pristine > F at chain > F at ring > F at chain and ring; while in graphdiyne, its order becomes pristine > F at ring > F at chain > F at chain and ring. It is also clear that for both the systems, the stability becomes less due to whole fluorination.

### *3.3 Density of states*

The Density of states for pristine graphyne and graphidyne are presented in supplementary material S3. We have found that the first state near the Fermi level is mainly contributed by $p_z$ orbitals of carbon atom which is consistent with our previous study [11].

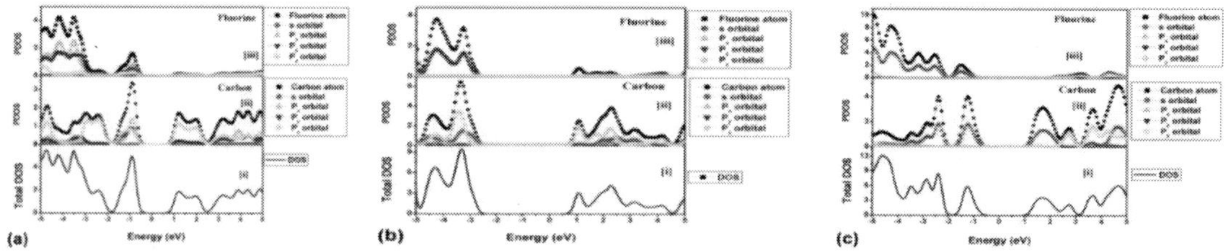


**Figure 3.** Total and partial DOS of structures: (a)graphyne; (b) graphyne with fluorine at chain; (c) graphyne with fluorine at ring; (d) graphyne with fluorine at both chain and ring.

In figures 3a, 3b and 3c, we plotted the total and partial density of states when graphyne is functionalized by F atoms at the linear chain site, at the hexagonal position and throughout the system respectively. The contribution of carbon and F atom is presented in figures 3a (ii) and 3a (iii). From figure3a (ii), it is clear that for carbon atom all the *p* orbitals is contributing in the valance band as well as in the conduction band. In case of fluorine atom, all the *p* orbitals are basically contributing (figure 3a (iii)) in the valence band only. For carbon atom, the $p_z$ orbital first start to contribute to the energy levels in the conduction band compare to other orbital whereas all the *p* orbitals starts to contributing in the valence band near the Fermi level.

Figures 3b (i), 3b (ii), and 3b (iii) show the total DOS and the contribution of individual atoms to the density of states when F atoms are placed in the hexagonal positions of graphyne. We observed that the carbon atom contributes in the conduction band as well as in the valence band, but the fluorine atom only contributes in the valence band. It is clear from figure 3b (ii) that all the *p* orbitals of C atoms (figure 3b (ii)) contribute both in the valance and conduction bands. We see that the energy level near the Fermi energy is mainly contributed by $p_z$ orbital of carbon atom in the conduction band while in the valence band the $p_x$ and $p_y$ orbital contribute first near the Fermi level. This feature was not seen when F atoms are situated at linear chain site of graphyne, in which all three *p* orbitals of C atoms contribute near the Fermi level in the valence band. The F atoms contribution is mainly coming from $p_x$ and $p_y$ orbitals in the valence band as can be seen from figure 3b (iii).

For graphyne with F at chain and ring (3c (i), 3c (ii) and 3c (iii)), the energy levels near the Fermi level of valence band and conduction band come from the $p_x$ and $p_y$ orbital of the carbon atoms which is not seen in case of graphyne with F atom at chain and graphyne with F atom at ring.

In this case due to whole fluorination the energy states contributed by $p_z$ orbital of carbon atoms are shifted towards high energy both in valence and conduction bands. Major contribution in the valence region comes from F atoms, but for conduction region, major contribution comes from C atoms. The same situation in the PDOS curve is observed for graphdiyne with F at chain, graphdiyne with F at ring and graphdiyne with F at chain and ring [see in supplementary material S3].

### *3.4 -COHP analysis*

The –COHP analysis for pristine graphyne, graphdiyne, are demonstrated in figures S4a, and S4b respectively [in supplementary material]. It is clearly visible that, in pristine case only C-C interaction exists and it contributes bonding state near the Fermi level.

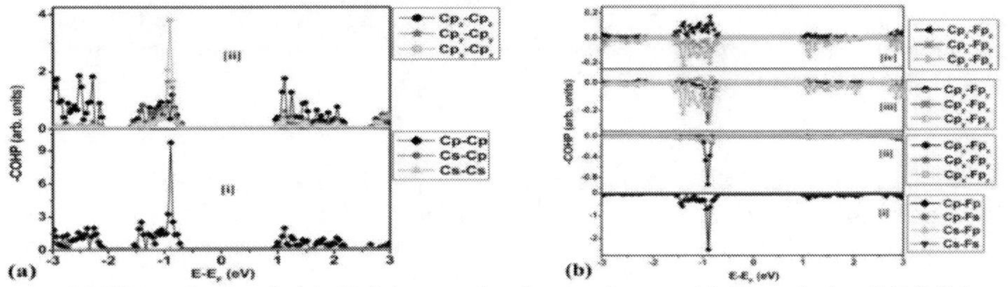


**Figure 4.** -COHP analysis of: (a) C-C interaction in graphyne with F at chain; (b) C-F interaction in graphyne with F at chain.

Then we report -COHP analysis by putting F atoms at chain (figure 4) and observed two considerable interaction, carbon-carbon (C-C) and carbon-fluorine (C-F). In C-C interaction (figure 4 (a)) of graphyne, contributed the bonding states near the Fermi level and all the bonding states are mainly contributed by *p-p* orbital pair. In the valence and conduction region near the Fermi level all the three *p* orbitals contribute. The C-F interaction (figure 4 (b)) mainly contributes antibonding state near the Fermi level the contribution arises mainly from *p-p* orbital pair.

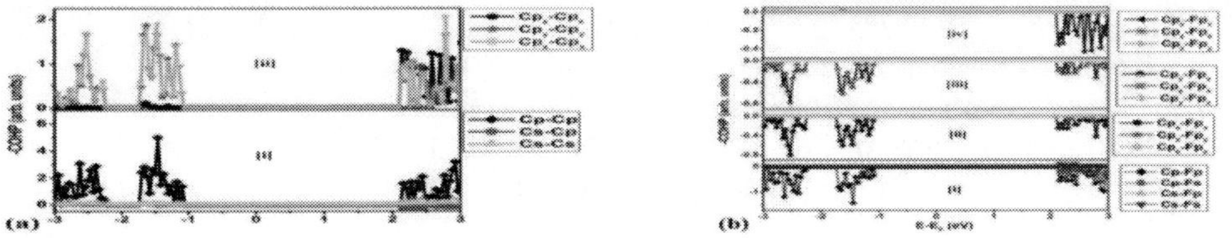

**Figure 5.** -COHP analysis of:(a) C-C interaction in graphyne with F at chain and ring; (b) C-F interaction in graphyne with F at chain and ring.

The -COHP analysis of graphyne family with F at ring (see in supplementary material S5) has a remarkable feature that, the presence of F at ring shifts the bonding and antibonding states away from the Fermi level (in both conduction band and valence band) than graphyne with F at chain site and thus increase the band gap. The C-C interaction contribute bonding state near the Fermi level both in valence and conduction region while C-F bond contribute antibonding state arises from *p-p*, and *s-p* orbital pair of carbon and fluorine atom.

Next we consider the –COHP analysis of graphyne with F atom at both chain and ring. In C-C interaction (figure 5 (a)) the bonding states near the Fermi level in the valence band mainly contributed by $p_x$-$p_x$ and $p_y$-$p_y$ pair of carbon atom and in CB all the three *p* orbitals contributes equally. The C-F interaction (figure 5 (b)) contributed antibonding state mainly dominated by *p-p* orbital pair of neighbouring carbon and fluorine atom. From the above analysis, it is clear that the functionalization by fluorine activate the participation of $p_x$-$p_x$, $p_y$-$p_y$, *s-p* and *s-s* orbital pair of neighbouring carbon atom near the Fermi level.

## 4. Conclusion

In summary, we have carried out density functional theory calculation of pristine and fluorinated graphyne and graphdiyne to investigate its electronic properties. It has been found that fluorination widens the band gap of graphyne and graphdiyne depending on its functionalization site showing the trend pristine < system with F at chain < systems with F at chain and ring < system with F at ring. The fluorination either at chain or at ring site remains these systems as direct band gap semiconductor, but due to full fluorination, these systems become an indirect semiconductor. Again the fluorination decrease the stability of these system and the whole fluorinated system shows less stability compare to other fluorinated system (system with F at chain or system with F at ring). The fluorine atom mainly contributed in the valence band and the fluorination at different sites of these fundamental structures activates the $p_x$, $p_y$, and *s* orbitals near the Fermi level. Our results show that, the C-C interaction of neighboring carbon basically contributes bonding state while the C-F interaction always contributes antibonding state near the Fermi level.

**Acknowledgement**

Dr U Sarkar would like to thank SHARCNET Canada for providing computational facilities for this research work and International Centre for Theoretical Physics (ICTP), Trieste, Italy for hosting him as a regular associate.

## References

[1] Novoselov KS, Geim AK, Morozov SV, Jiang D, Zhang Y, Dubonos SV, Grigorieva I V and Firsov A A 2004 *Science* **306** 666
[2] Zhang Y, Tan Y W, Stormer H L and Kim P 2005 *Nature* **438** 201
[3] Hirsh A 2010 *Nat. Mater.* **9** 868
[4] Katsnelson MI, Novoselov KS and Geim A K 2006 *Nat. Phys.* **2** 620
[5] Balog R, Jørgensen B, Nilsson L, Andersen M, Rienks E, Bianchi M, Fanetti M, Lægsgaard B, Baraldi A, Lizzit S, Sljivancanin Z, Besenbacher F,Hammer B, Pedersen T G, Hofmann P and Hornekær L 2010 *Nat. Mater.* **9** 315
[6] Jeon K J, Lee Z, Pollak E, Moreschini L, Bostwick A, Park C M, Mendelsberg R, Radmilovic V, Kostecki R, Richardson T J and Rotenberg E 2011 *ACS Nano* **5** 1042
[7] Lin Y, Ding F and Yakobson B I 2008 *Phys. Rev. B* **78** 041402.
[8] Malko D, Neiss C, Vin˜es F, Go¨rling F 2012 *Phys. Rev. Lett.* **108** 086804
[9] Kehoe J M, Kiley J H, English J J, Johnson C A, Petersen R C and Haley M M 2000 *Org. Lett.* **2** 969
[10] Li G X, Li Y L, Liu H B, Guo Y B, Li Y J and Zhu D B 2010 *Chem. Commun.* **46** 3256
[11] Singh N B, Bhattacharya B and Sarkar U Struct Chem, DOI 10.1007/s11224-014-0440-4
[12] Oda T, Pasquarello A and Car R 1998 *Phys. Rev. Lett.* **80** 3622
[13] Soler J M, Artacho E, Gale J D, García A, Junquera J, Ordejón P and Portal D S 2002 *J. Phys: Condens. Matter.* **14** 2745
[14] Troullier N and Martins J 1990 *Solid. State. Commun.***74** 613
[15] Kang J, Li J, Wu F, Li SS and Xia J B 2011 *J. Phys. Chem. C* **115** 20466

# Plasmonic Structures for Sensing and Emitting Devices

**Francesco Floris[1], Lucia Fornasari[1], Maddalena Patrini[1], Cristiana Figus[2], Andrea Mura[2], Giovanni Bongiovanni[2], Francesco Quochi[2], Paola Pellacani[3], Andrea Valsesia[3] and Franco Marabelli[1]**

[1] Department of Physics, University of Pavia, Via Agostino Bassi 6, 27100 Pavia (PV), Italy
[2] Department of Physics, University of Cagliari, Complesso Universitario di Monserrato S.P. Monserrato-Sestu Km 0,7, 09042 Monserrato (CA), Italy
[3] Plasmore S.r.l. – Via Grazia Deledda 4, 21020 Ranco (VA), Italy.

Corresponding author's e-mail address: francesco.floris01@ateneopv.it

**Abstract**. We report on the study of a plasmonic nanostructure that could be adopted as platform for emitting and sensing applications. Several devices have been prepared and characterized by atomic force microscopy (AFM) and Fourier transform micro-reflectance (FT-μR) techniques. In addition, a modelling via finite-difference time-domain (FDTD) simulations have been developed in order to interpret the morphological shape and the optical response of the considered structures. Until now, remarkable performances as surface plasmon resonance (SPR) based optical sensor have been founded. Moreover, we are performing preliminary trials in order to establish a coupling between photoluminescence (PL) features of suitable emitters with respect to the plasmonic resonances.

## 1. Introduction

The interaction between light and matter is one of the most fundamental topics in physics.

In most cases, metals are just considered materials for realizing reflectors. However, there is a useful light–matter interaction involving metals that enables us to use metals in a more refined manner. Surface plasmons (SPs) are electromagnetic (E.M.) modes involving mobile surface charges, typically the conduction electrons in metals.

Such charges react to an external E.M. field oscillating in a coherent collective way, exhibiting a behavior of a vast ensemble of interacting particles. Under appropriate energy and momentum matching conditions, light interact with SPs and absorption and excitation processes can be observed.

SPs modes have greater momentum than light of the same frequency and, as a consequence, the E.M. fields associated with this oscillations cannot propagate away from the surface: the wave is evanescent and the field strength decays exponentially with increasing distance from the surface.

SPs modes on a planar metal surface are thus bound to that surface and guided by it, propagating until their energy is dissipated as heat in the metal.

From this point of view, plasmonics represents a way to squeeze electromagnetic fields below the diffraction limit at a metal-dielectric interface, with a consequential huge confinement of the corresponding E.M. energy [1-4].

Then, in order to realize a working device, the probes (in the case of sensing applications) or the fluorophores (in the case of emitting applications) must be placed in the near-field region of the contact surface between the metal and the dielectric, within the decay length (a few tens of nanometers) of the plasmonic mode [5].

To this aim, we planned a morphology able to ensure an optimal spatial positioning of the plasmonic resonator, with an accessible and large contact area between the metal and the external environment. These properties allow us to apply this structure both in sensing and emitting applications.

## 2. The Plasmonic Structure

We made use of a plasmonic surface consisting of a gold film embedded in a two-dimensional array of dielectric polymeric pillars [6-8].

Figure 1 illustrates the design of a single pillar structure and their regular repetition compose the plasmonic nanostructure. It is also visible the shape of the main plasmonic resonance field distribution (placed at a wavelength of ~820 nm) in the real space. The extension of the field toward the external environment is the key to make the plasmonic field usable for applied implementations.

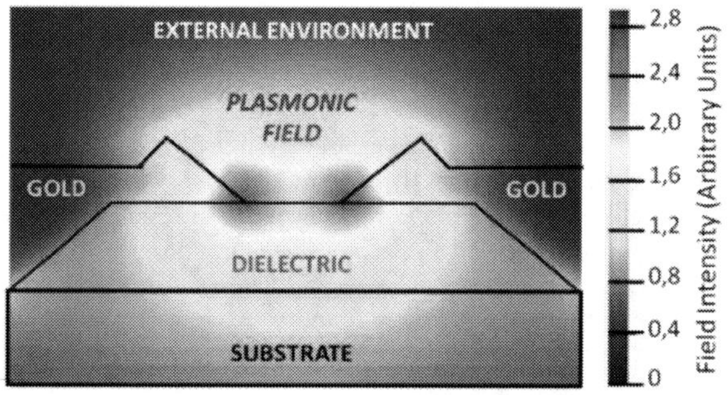


**Figure 1.** Electric field spatial distribution simulated using the finite-difference time-domain (FDTD) numerical analysis technique. The simulation is performed in the cross section of the pillars at the wavelength of the main spectral components of the plasmonic resonance (~820 nm). The superimposed scheme shows the various sections that compose the nanostructure.

Pillars are arranged in a triangular array in order to obtain a hexagonal lattice of plasmonic resonators (Fig. 2). With this particular configuration, we obtained an ordered series of localized resonators, circularly shaped and close enough to establish reciprocal coupling.

In this way, we have an E.M. network, confined at the surface of contact between the plasmonic nanostructure and the external environment, made of a combination of localized and travelling evanescent waves. Such a net is providing a high sensitivity on changes of the surrounding dielectric function that make the whole system perfect for sensing applications. Vice versa, thanks to its extensiveness, both in the surface plane and in the transversal one, the field mesh is favouring a good coupling with suitable fluorophores laid down on its top. Moreover, it should be possible to exploit the plasmonic field for pumping the fluorophores in order to sustain their photoluminescence (PL) features for emitting applications.

Journal of Physics: Conference Series **566** (2014) 012015 doi:10.1088/1742-6596/566/1/012015

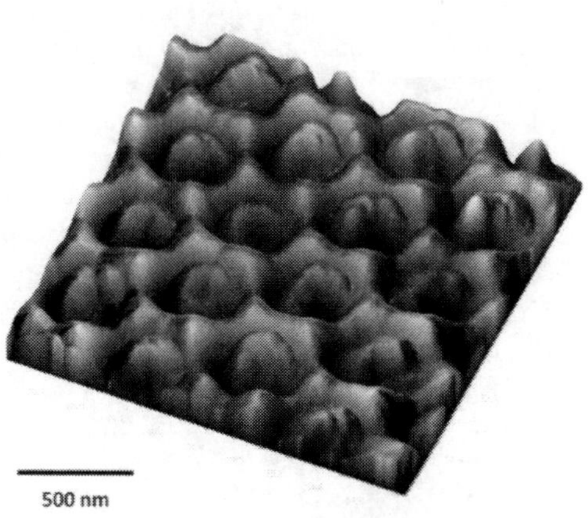


**Figure 2.** Three dimensional virtual image of the surface of the plasmonic structure obtained by an AFM measurement. The triangular array of pillars is well visible as the corresponding hexagonal lattice of plasmonic resonators.

## 3. Sensing Applications

The main feature of the optical response in the near infrared of the plasmonic surface can be interpreted in terms of plasmonic excitations localized at the contact edge among the head of the dielectric pillars, the gold slab and the free external surface.

Due to the high sensitivity of the plasmonic surface to refractive index $n$ variations, any alteration in the chemical and physical composition of the external environment affect the spectral position of the plasmonic resonance.

The detection of such a refractive index change in the external environment is obtained by measuring the reflectance spectra with light incident from its substrate side, before and after the change (Fig. 3 shows an example).

The main resonance is shown as a reflectance minimum at ~820 nm.

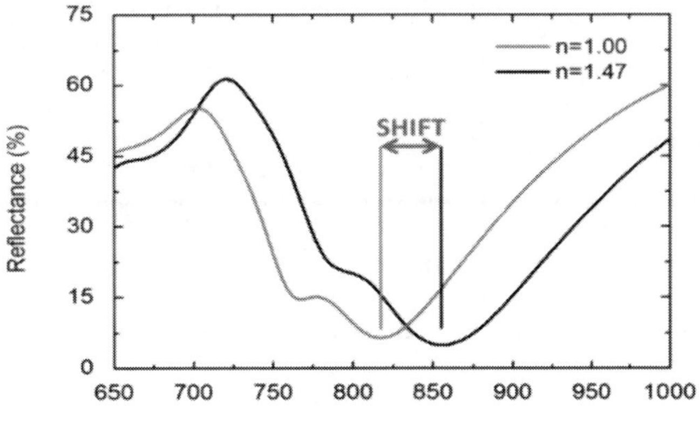


**Figure 3.** Reflectance spectrum between 650 and 1000 nm measured with TM polarized light at nearly normal incidence. In this particular case, the plasmonic resonance shift is due to the deposition of a thin (~10 nm) silicate layer (refractive index = 1.47) over the device surface.

Operatively, the plasmonic response can be evaluated as the ratio among reflectance spectra (Fig. 4-a). It has been shown that the numerical value of this ratio is directly proportional, at each wavelength, to the spectral shift induced by the variations in the external environment refractive index *n* [9] and the consequent sigmoid-like function can be used to evaluate the corresponding refractive index variation signal amplitude. Several measurements with ethanol, sodium chloride, glycerol and polyelectrolytes were performed in order to prove the surface capability. Further experiments have been realized to apply the nanostructure for medical diagnostic purpose [9].

Moreover, a combined study of sol-gel synthesis and dip-coating method, enabled us to deposit well-controlled ultrathin silica layers on the plasmonic nanostructure. Such layers allow both surface protection and morphology preservation and, in addition, it has been proved that, for ultrathin silica thickness, they establish sensitivity gain in detecting refractive index changes (Fig. 4-b) [10].

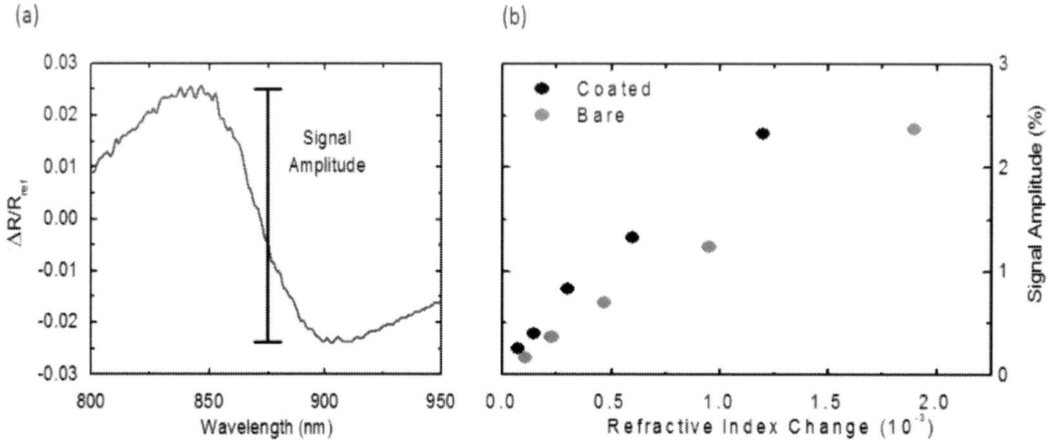


**Figure 4.** (a) Ratio of two reflectance measurements. ΔR is equal to the difference between the reflectance spectrum after the refractive index change (in this example a mixture of milli-Q water and ethanol at 3%) and the reflectance reference spectrum, $R_{ref}$, before the refractive index modification (milli-Q water only). (b) Plasmonic response to several ethanol solutions with different concentrations for the bare (black circles) and the ultrathin silica coated (red circles) surface. The plot displays the amplitude changes in reflectance referred to the signal shown in (a). Errors are very small (the maximum error for the refractive index change is $2x10^{-5}$ and the maximum error for the signal amplitude is 0.05%) and the corresponding error bars would not be clearly visible in the picture.

## 4. Emitting Applications

Plasmonic-photonic coupled devices, promoting the interaction between surface plasmons based devices and luminescent media, are very promising for nano-photonic applications [3].

In principle, the high electromagnetic density of the plasmonic field should be able to feed the fluorescence emission of suitable layers deposited within the field extension.

In this way, we realized preliminary trials in order to promote the coupling between PL emission of a standard fluorophore (rhodamine R6G) with respect to the plasmonic resonance exhibited by our nanostructure.

Until now, we focused our attention on the existence of a useful way to deposit the emitters on the nanostructure. We paid particular attention to avoid photoluminescence quenching effects due to metal absorption (gold in our specific case) when a fluorophore, is in close proximity with it.

Therefore, we coated the nanostructured surface with a thin film layer, ~10 nm in thickness, made of silica ($SiO_2$). Such *white-gauge* carries out the task of avoid the direct contact between gold and rhodamine. Anyway, its tiny width ensures a positioning of rhodamine within the plasmonic field penetration depth. The result is an efficient attachment platform for fluorophore molecules.

Consequently, we deposited the rhodamine by post-grafting on both the surfaces, the bare and the coated, and proved the quenching induced by the gold-R6G direct contact on the fluorophore PL emission together with the benefit introduced in the system by the silica coating.

The PL results are reported in figure 5.

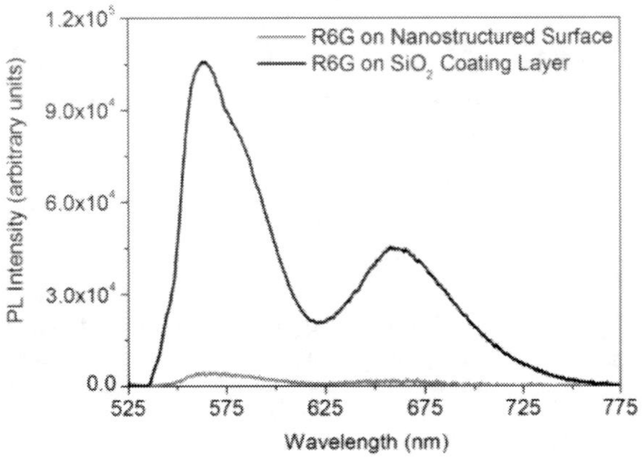


**Figure 5.** Photoluminescence signal of rhodamine (R6G) dropped off by post-grafting directly on the nanostructured surface (red line) and on the silicate coating layer (black line). The effect of quenching on the PL intensity emission, induced by the gold when it is in contact with the fluorophore, is well-rendered. Both the systems were excited with a laser source at the pumping wavelength of 532 nm.

## 5. Conclusions

In summary, we developed a platform consisting in a plasmonic nano-structured surface based on a gold film embedding a two-dimensional array of polymeric pillars, able to be adopted as platform for emitting and sensing applications. Numerous devices have been realized and characterized by atomic force microscopy, Fourier transform micro-reflectance and spectroscopic ellipsometry techniques.

Significant performance as surface plasmon resonance based optical sensor, for both biological and medical diagnostic applications, have been proved. In addition, we realized encouraging attempts to verify the likely occurrence of a coupling between photoluminescence emission of a standard fluorophore (rhodamine) with the plasmonic resonance supported by our nanostructure.

## References

[1] W. L. Barnes, *J. Opt. A: Pure Appl. Opt.* **8**, S87-S93 (2006)
[2] S. A. Maierm in *Plasmonics: Fundamentals and Applications*, Springer (2007)
[3] R. Ma, R. F. Oulton, V. J. Sorger, X. Zhang, *Laser Photonics Rev.* **7**, 1-21 (2013)
[4] M. E. Stewart, C. R. Anderton, L. B. Thompson, J. Maria, S. K. Gray, J. A. Rogers, R. G. Nuzzo, *Chem Rev.* **108**, 494-521 (2008)
[5] W. L. Barnes, A. Dereux, T. W. Ebbesen, *Nature* **424**, 824–830 (2003)
[6] S. Giudicatti, A. Valsesia, F. Marabelli, P. Colpo, F. Rossi, *Phys. Status Solidi A* **207**(4), 935–942 (2010)
[7] S. Giudicatti, F. Marabelli, A. Valsesia, P. Pellacani, P. Colpo, F. Rossi, *J. of Optical Society of America B* **29**, 1641-1647 (2012)
[8] S. Giudicatti, F. Marabelli, P. Pellacani, *Plasmonics* **8**(2), 975-981 (2013)
[9] B. Bottazzi, L. Fornasari, A. Frangolho, S. Giudicatti, A. Mantovani, F. Marabelli, G. Gerardo, P. Pellacani, R. Therisod, A. Valsesia, *J. of Biomedical Optics* **19**(1), 017006 (2014)
[10] F. Floris, C. Figus, L. Fornasari, M. Patrini, P. Pellacani, G. Marchesini, A. Valsesia, F. Artizzu, D. Marongiu, M. Saba, A. Mura, G. Bongiovanni, F. Marabelli, F. Quochi, *J. Phys. Chem. Lett.* **5**, pp 2935–2940 (2014)

Journal of Physics: Conference Series **566** (2014) 012016 doi:10.1088/1742-6596/566/1/012016

# Structure and gap of low-$x$ $(Ga_{1-x}In_x)_2O_3$ alloys

**M B Maccioni, F Ricci and V Fiorentini**

CNR-IOM and Dept. of Physics, University of Cagliari, Cittadella Universitaria, 09042 Monserrato (CA), Italy

E-mail: mariabarbara.maccioni@dsf.unica.it, vincenzo.fiorentini@dsf.unica.it

**Abstract.** We study the electronic and local structural properties of pure and In-substituted $\beta$-$Ga_2O_3$ using density functional theory. Our main result is that the structural energetics of In in $Ga_2O_3$ causes most sites to be essentially inaccessible to In substitution, thus limiting the maximum In content to somewhere between 12 and 25 % in this phase. We also find that the band gap variation with doping is essentially due to "chemical pressure", i.e. volume variations with doping.

## 1. Introduction

$Ga_2O_3$ is attracting interest recently as a material for high-power transport and ultraviolet optical absorbers, owing to its wider band gap and larger electric breakdown voltage compared to e.g. GaN. Combined with $In_2O_3$ (already widely used as transparent conducting oxide), $Ga_2O_3$ originates a new materials system which is tunably insulating, easily $n$-doped (not so easily $p$-doped), and potentially magnetic (as $In_2O_3$ can be made ferromagnetic [1] through magnetic doping, the same may well apply to $Ga_2O_3$). Further, the band-engineering and nanostructuration concepts from popular semiconductor systems such as, e.g., AlGaAs or InGaN may be exported to these materials, and thus to a whole new region of high absorption energies and breakdown voltages. This may enable the design of devices based on $Ga_2O_3/(Ga_{1-x}In_x)_2O_3$ such as high-power field effect transistors and far-UV photodetectors or emitters.

## 2. $Ga_2O_3$

Gallium oxide, $Ga_2O_3$, exists in various polymorphs, the most stable being monoclinic $\beta$-$Ga_2O_3$ at ambient condition [2]. The monoclinic phase, shown in Figure 1, belongs to the C2/m space group. The unit cell contains 20 atoms, with two crystallographically nonequivalent Ga atoms in tetrahedral and octahedral like coordinations in the lattice. Geometry and volume optimizations as well as electronic structure calculations have been done using density-functional theory (DFT) in the generalized gradient approximation (GGA), and the Projector Augmented-Wave (PAW) method as implemented in the VASP code [3]. The bulk Brillouin zone is sampled on a 4×8×6 Monkhorst-Pack grid. The calculated lattice parameters compare well with experiment [4] (in parenthesis): $a$=12.46 Å (12.23), $b$=3.08 Å (3.04), $c$=5.88 Å (5.80), $\theta$=103.65° (103.7).

The band structure of $Ga_2O_3$, similarly to other binary Ga compounds, has mainly O $2p$ character in the upper valence band and Ga $s$ content in the bottom conduction band. The direct gap appears at the Γ point. GGA underestimates it at about 2 eV, as expected of semilocal functionals. Adding an empirical self-energy correction [5] involving the high-frequency dielectric constant, we obtain a gap of 4.2 eV, not far from the experimental range of 4.5-5 eV [6, 7].

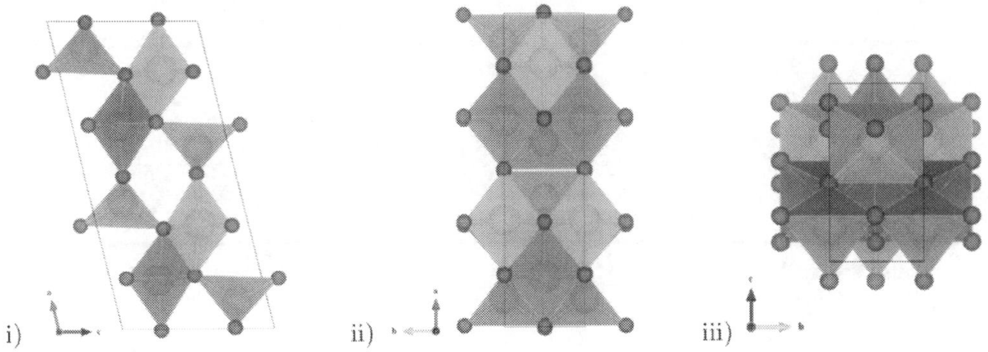

**Figure 1.** Monoclinic $\beta$-$Ga_2O_3$ viewed along the b-axis (i), c-axis (ii) and a-axis (iii). The structure can be seen as a collection of zigzag double chains of edge-sharing $GaO_6$ units (violet and green Ga-octahedra) linked by single chains of vertex-sharing $GaO_4$ (blue Ga-tetrahedra).

Surprisingly, the precise value of the gap is still uncertain even in recent work [7]. The reason for this is probably the significant anisotropy of the absorption, which we have analyzed (and will report elsewhere [8]) with hybrid-functional, pseudo self-interaction corrected, and $GW_0$ many-body perturbation-theory calculations (known to be free of the typical LDA/GGA gap errors). For the present purposes, we just note that these advanced methods confirm that a direct minimum gap at zone center between 4.2 eV (hybrids) and 4.7 eV (self-interaction-correction), and also confirm the pressure derivative of the gap to be 3 meV/kbar essentially as in GGA (see below).

## 3. Low-In-content alloying

Because unalloyed In and Ga oxides have different structures (bixbyite and monoclinic $\beta$, respectively) the high-In and low-In-content alloying limits will behave quite differently, and at intermediate concentrations the two phases are likely to mix in an complicated way. The experimental alloying of $Ga_2O_3$ with $In_2O_3$ indeed faces significant limitations [7], with $\beta$-$Ga_2O_3$-like and bixbyite-like X-ray spectra at low $x$ and high $x$ respectively, and a mixed-phase region at midrange $x$. In particular the $\beta$-$Ga_2O_3$-like phase persists only up to about 15% or so [7]. Thus, keeping in mind that the large-$x$ end of the alloying spectrum will have to be treated differently, here we tackle the low-$x$ end substituting In for Ga in $\beta$-$Ga_2O_3$ at nominal concentrations of 3, 6, 9, and 12 % (one to four In atoms per 80-atom or 32-cation supercell). Our results naturally suggest an interpretation for the observed behavior.

We optimize (in volume, shape, and internal coordinates) supercells of $Ga_2O_3$ with In$\rightarrow$Ga substitutions sampling some of the various possible octahedral and tetrahedral sites and combinations thereof as function of composition (i.e. In concentration). The supercells contain $1\times2\times2$ unit cells (80 atoms), so we an approximate backfolding ($4\times4\times2$) of the bulk k-point grid. The calculations at 3% In (one "isolated" In atom per 80-atom cell) show that In only substitutes octahedral Ga: tetrahedral sites are ruled out by a huge excess energy cost of 1.1 eV. Therefore, half the cation sites are essentially inaccessible to In, and hence the amount of In that can actually be incorporated into Ga oxide is automatically halved – to put it differently, all available sites would be occupied already at 50% nominal In content.

Even reaching this limit, though, is quite unlikely. Indeed, In incorporation is not arbitrary in terms of configuration. At 6% In content, i.e two In atoms per supercell, one can estimate the energetics of pairing (or, rather, non-pairing) of In in $Ga_2O_3$. In Figure 2, left, we display the preferential couplings. The energy of configuration '0' is chosen as zero; the structures numbered

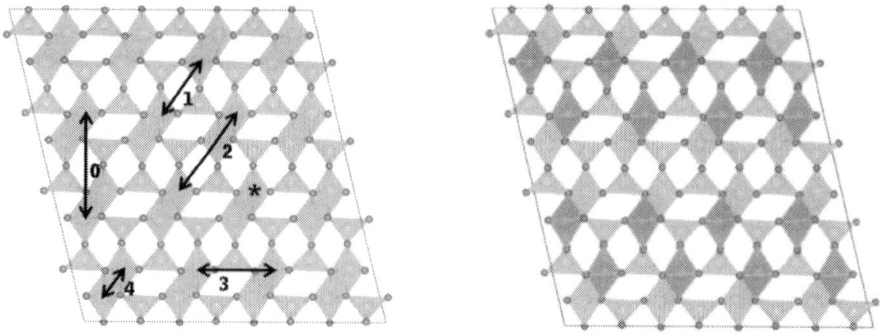


**Figure 2.** Left: different pairings patterns of In in $Ga_2O_3$. Right: structure of the 12% sample.

'1' to '4' are in progressively unfavorable energetic order, with '1' at 16 meV, '2' at 50 meV, '3' at 100 meV, '4' at 125 meV. In the configuration labeled '*', In atoms occupy adjacent octahedra; this structure is 250 meV higher than the '0'-configuration.

Clearly, In atoms tend to avoid one another, and it is likely that at the common growth temperatures of 850 K the typical configurations will be such as our '0' and '1' above. Inspecting the structure, this suggests that well below a half, and probably closer to a quarter, of the octahedral sites can be occupied by In with a reasonable energy cost; when these are filled, the formation of some mixed $\beta$/bixbyite phase may be preferable to substitution in the $\beta$ phase. This brings the effective solubility in the original $\beta$-$Ga_2O_3$ structure down to between 12% and 25% as found in experiment [7].

Consistently with the above configurational restrictions on pairing, the admissible arrangements at 9% and especially 12% In content are few. For example, the structure used at 12% is in Figure 2, right. The resulting structures are probably a fairly decent model of the alloy, given the very limited configurational freedom of In already at these concentrations. (Cluster-expansion work is ongoing on miscibility at finite temperature and will be reported elsewhere.)

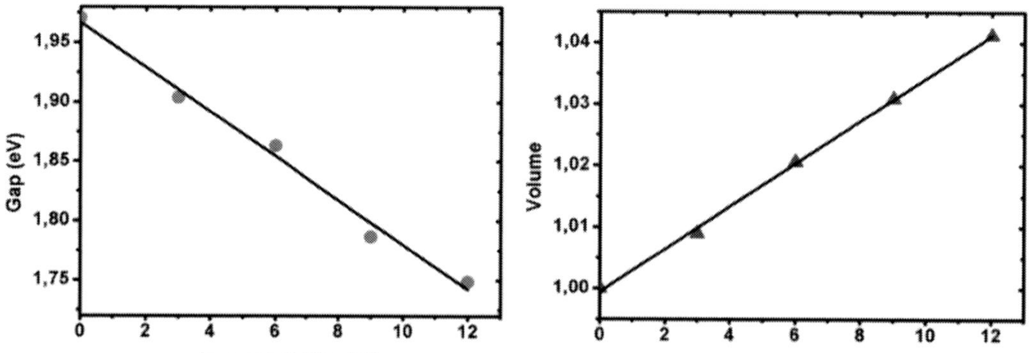


**Figure 3.** Left: direct gap at zone center as function of In concentration. Right: normalized volume of In-doped $Ga_2O_3$ as function of concentration.

We calculated the gap and the volume change with concentration of $(Ga_{1-x}In_x)_2O_3$ in the 80-atom supercell for the structures mentioned. The gap is in Figure 3, left, and the volume is in the same Figure, right. Although the gap is underestimated (a token of using GGA), its concentration change of 17 meV/% is quite similar to 20 meV/% experimentally [7]. Using the the volume change with $x$ and the bulk modulus, we can evaluate the concentration derivative as a pressure derivative, obtaining 2.5 meV/kbar. This is similar to the $Ga_2O_3$ value of 3 meV/kbar [8, 9], which suggests that the gap is mainly affected by volume change, and marginally by other factors. In this light, the agreement with experiment therefore falls in line with expectations from previous work [10].

## 4. Conclusions

In summary, we have performed first-principles DFT-GGA calculations on the electronic and local structural properties of In-containing $Ga_2O_3$. The energetics of In in $Ga_2O_3$ limits the maximum In content to somewhere between 12 and 25 %. We also find that the gap variation with doping is essentially due to "chemical pressure", i.e. volume variations with doping.

## Acknowledgments

M.B.M. gratefully acknowledges Sardinia Regional Government for the financial support of her PhD scholarship (P.O.R. Sardegna F.S.E. Operational Programme of the Autonomous Region of Sardinia, European Social Fund 2007-2013 - Axis IV Human Resources, Objective l.3, Line of Activity l.3.1). This work was also supported in part by MIUR-PRIN 2010 project *Oxide*, Fondazione Banco di Sardegna and CINECA grants.

## References

[1] Alippi P, Cesaria M, and Fiorentini V 2014 *Phys. Rev. B* **89** 134423.
[2] Edwards D F 1998 *Handbook of Optical Constants of Solids* Academic Press, **III** 753
[3] Kresse G and Furthmuller J 1996 *Phys. Rev. B* **54** 11169
[4] Geller S 1960 *J. Chem. Phys.* **33** 676
[5] Fiorentini V and Baldereschi A 1992 *J. Phys.: Condens. Matter* **4** 5967
[6] Tippins H H 1965 *Phys. Rev. B* **140** 316
[7] Zhang F et al 2014 *Solid State Comm.* **186** 28
[8] Ricci F et al 2014, to be published.
[9] He H et al 2006 *Phys. Rev. B* **74** 195123.
[10] Fiorentini V 1992 *Phys. Rev. B* **46** 2086

Journal of Physics: Conference Series **566** (2014) 012017 doi:10.1088/1742-6596/566/1/012017

# Colloidal synthesis and characterization of $Bi_2S_3$ nanoparticles for photovoltaic applications

**R Piras[1], M Aresti[1], M Saba[1], D Marongiu[1], G Mula[1], F Quochi[1], A Mura[1], C Cannas[2], M Mureddu[2], A Ardu[2], G Ennas[2], V Calzia[3], A Mattoni[3], A Musinu[2] and G Bongiovanni[1]**

[1] Dept. of Physics
[2] Dept. of Chemistry
[3] CNR-IOM, University of Cagliari, Cittadella Universitaria, 09042 Monserrato (CA), Italy

E-mail: roberto.piras@dsf.unica.it, michele.saba@dsf.unica.it, musinu@unica.it

**Abstract.** Bismuth sulfide is a promising n-type semiconductor for solar energy conversion. We have explored the colloidal synthesis of $Bi_2S_3$ nanocrystals, with the aim of employing them in the fabrication of solution-processable solar cells and to replace toxic heavy metals chalcogenides like PbS or CdS, that are currently employed in such devices. We compare different methods to obtain $Bi_2S_3$ colloidal quantum dots, including the use of environmentally benign reactants, through organometallic synthesis. Different sizes and shapes were obtained according to the synthesis parameters and the growth process has been rationalized by comparing the predicted morphology with systematic physical-chemistry characterization of nanocrystals by X-ray diffraction, FT-IR spectroscopy, Transmission Electron Microscopy (TEM).

## 1. Introduction

Colloidal semiconductor nanocrystals are promising materials for applications in solution-processable optoelectronic devices such as photocatalysts and photovoltaic cells. A wide variety of materials has been used to fabricate semiconductor nanocrystals, but most of them contain heavy metal elements, such as Pb or Cd, which could be dangerous for the health if used on large scale. For this reason, environmentally benign nanocrystals are currently a central topic in materials research. Bismuth sulfide ($Bi_2S_3$) belongs to the class of non-toxic semiconductor materials that can be synthesized in nanocrystalline form by colloidal synthesis; research efforts have been addressed to the fabrication of solution-processed optoelectronic devices, including hybrid bulk nano-heterojunctions for solar energy conversion and sensitized solar cells [1, 2, 4]. The synthesis reported in literature with the best control of nanocrystals size and shape, as well as resulting in the highest conversion efficiencies for solar cells, employ however dangerous chemicals, such hexamethyldisilathiane (HMS), and do not appear sustainable for large scale productions [1, 2]. In this paper we compare several synthesis methods to obtain bismuth sulfide nanoparticles in different shape and morphology, maintaining the same crystallographic phase; particularly, we develop "green" approaches and benchmark them against the well-established techniques extensively reported in literature.

Journal of Physics: Conference Series **566** (2014) 012017 doi:10.1088/1742-6596/566/1/012017

## 2. Synthesis and results

**$Bi_2S_3$-HMS.** The standard route to produce $Bi_2S_3$ colloidal nanocrystals presented in literature is an oleic acid-based organometallic synthesis. A mixture of bismuth(III) acetate, oleic acid (OA) and 1-octadecene (ODE) was heated under stirring in argon atmosphere at 90 °C for 16 h and then heated up to 170 °C. A solution of hexamethyldisilathiane (HMS) in ODE was quickly injected into the flask at the same temperature (170°C) or after cooling down at 100°C or at 50°C. The samples were cooled down to room temperature and the nanoparticles were purified by successive dispersion/reprecipitation and centrifugation steps in toluene/methanol and finally dispersed in anhydrous toluene [1, 2]. These dispersions are colloidally stable for months. The HMS can easily produce $H_2S$, a dangerous, toxic gas. This means that this synthesis is not safe to scale up for large productions. Moreover this synthesis is very sensitive to little variation of its atmosphere and this produces damage on the stability of the colloidal dispersion. For this reason we have tried the following methods, using elemental sulfur as precursor in different solvents.
**$Bi_2S_3$-ODE.** This synthesis is equal to the $Bi_2S_3$-HMS except for the use of S precursor, that now is a solution of elemental S heated at 100°C for 20 min and then at 180°C for 1 h in ODE [2, 3]. This method needs a temperature growth of 170°C to produce a crystalline sample.
**$Bi_2S_3$-OAm.** In this case the sulfur precursor was a sulfur/oleylamine solution, produced by vigorous stirring at room temperature of the sulfur powder in oleylamine (OAm), to obtain a red clear appearance. The same mixture of bismuth acetate, oleic acid and 1-octadecene was heated under stirring at 90°C in vacuum for 1 h, and then heated up at 170°C in argon atmosphere. The solution was quickly injected into the flask. After 10 min, 30 min or 6 h, the reaction was quenched with ethanol and purified by dispersion/reprecipitation and centrifugation steps with toluene/ethanol. At the end, the sample was collected in toluene [4]. This method was tested in excess (ExS), in defect (DfS) of sulfur or with Bi/S stoichiometric ratio (StS).
**$Bi_2S_3$-NW.** This synthesis is similar to the $Bi_2S_3$-OAm except for the fact that the ODE and OA mixture was replaced by pure oleylamine. A mixture of bismuth acetate and OAm was stirred and heated to 130°C over the course of 30 min and then kept at the same temperature for 30 min, under Ar atmosphere. The solution of S/OAm was quickly injected in the flask. After 30 min at 100°C the reaction was quenched with cold toluene. Unsolubilized materials were removed by centrifugation, and acetone was added to the supernatant until it became turbid. The mixture was centrifuged, the supernatant discarded and the precipitated nanocrystals redispersed in toluene [5].

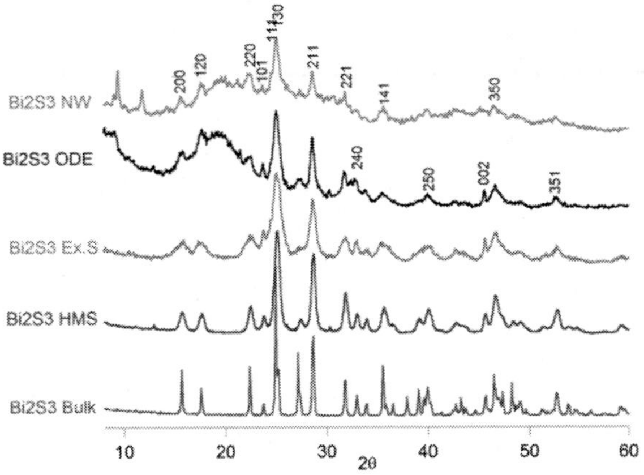


**Figure 1.** XRD patterns of NW, ODE, OAm ExS, HMS samples compared with the $Bi_2S_3$ bulk pattern.

X-ray diffraction (XRD) pattern of the most representative $Bi_2S_3$-HMS sample obtained under stoichiometric conditions after 6 h of growth, is reported in Figure 1, compared with the XRD patterns of the other samples and bulk bismuth sulfide. Bragg reflections indicate the presence of $Bi_2S_3$ with orthorhombic structure as the unique phase in the sample. All the other synthesis methods we employed produce bismuth sulfide in orthorhombic phase, like the $Bi_2S_3$-HMS, but with a different morphology.

TEM bright and dark field images of the $Bi_2S_3$-HMS sample at temperature of growth (Tg) of 170°C, reported in Figure 2, show the formation of rod shaped nanoparticles of average length of 25 nm (10% polydispersity) and width of 13 nm (13% polydispersity). The brightness of the nanorods in the dark field image is evidence that most of them are single crystals. Changing the temperature of growth, is possible to control size and shape of the nanocrystals. With lower temperature, in fact, the nanoparticles change from rod shape to quasi-spherical shape with a size of 4 nm.

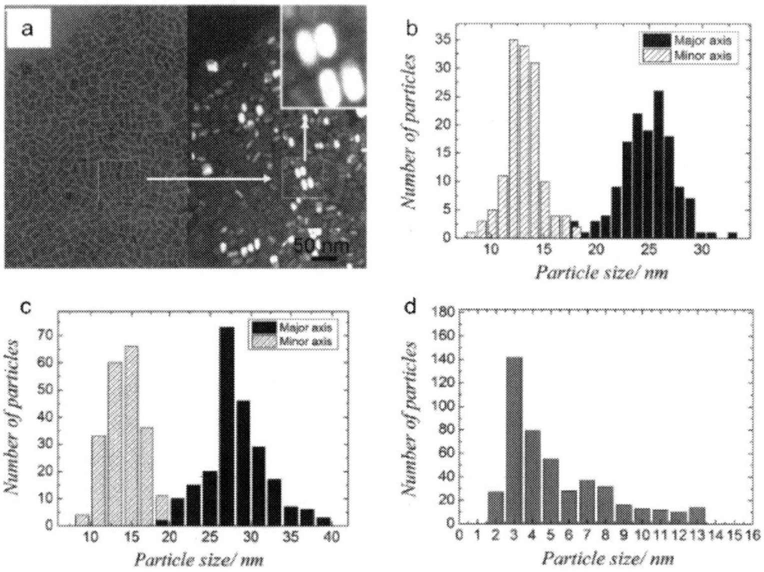


**Figure 2.** (a) TEM bright and dark field images of Tg = 170°C $Bi_2S_3$-HMS sample and is (b) particle size distributions calculated by bright field images. (c) Particle size distributions of Tg = 100°C sample. (d) Particle size distributions of Tg = 50°C sample [1].

In the Figure 3 are shown the TEM images of the most representative sample obtained with the other methods, compared with the $Bi_2S_3$-HMS sample. The ODE sample is a petal-like nanoparticle formed by needles of sulfide. The average length is 300 nm and these particles are not stable in organic solvent, so it is not a solution-processable sample. The OAm sample is amorphous if synthesized with stoichiometric ratio, very little in sulphur defect and rod-shaped in sulfur excess. This last one grows after 10 min at the same average length of the HMS sample after 6h of growth, but the width is halved. The time of growth influences the stability of this sample; if this is higher than 1 h, the yield of stable nanoparticles decrease. The $Bi_2S_3$-NW sample is formed by nanowires with a average width of 3-4 nm and a length over 300 nm. These ones are very stable in toluene. The preferential growth along a single dimension is attributable to the absence of oleic acid, which then has a primary role in the growth in the spherical form of the bismuth sulfide.

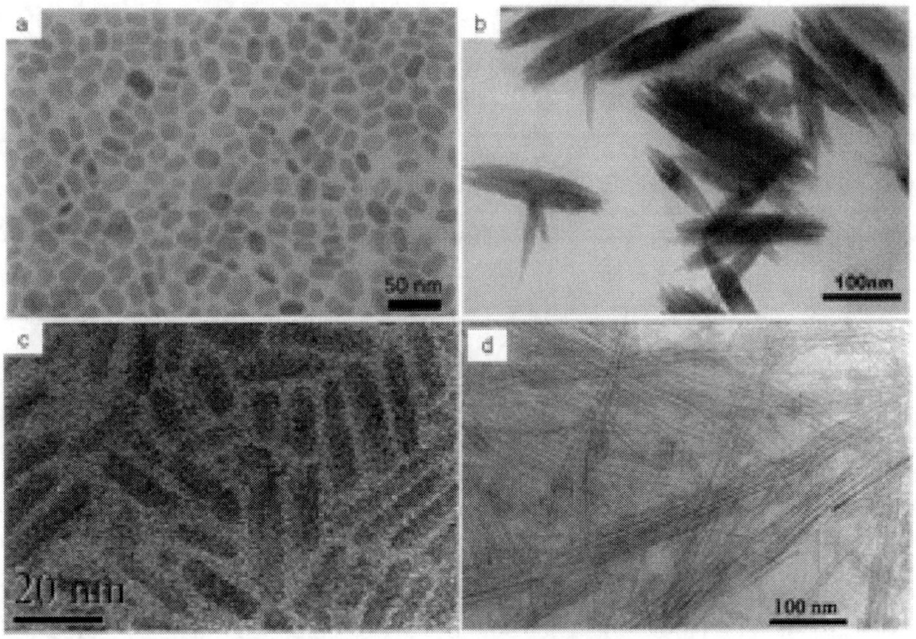


**Figure 3.** TEM images of (a) $Bi_2S_3$-HMS sample, (b) $Bi_2S_3$-ODE sample and HR-TEM images of (c) $Bi_2S_3$-OAm ExS and (d) $Bi_2S_3$-NW.

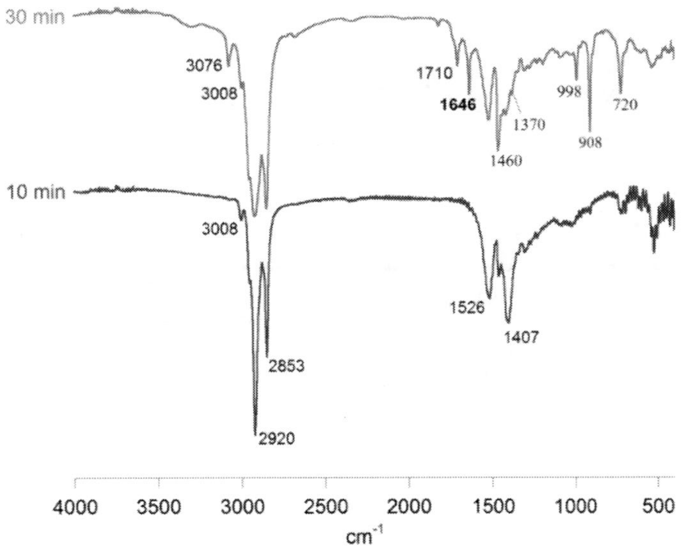


**Figure 4.** FT-IR spectra of $Bi_2S_3$-OAm sample after 10 min and 30 min of growth.

The correlation between the time of growth of the particles, after the injection of the sulfur solution, and its stability could be explained with the help of the infrared spectroscopy. The FT-IR spectra of the $Bi_2S_3$-OAm sample, are shown in the Figure 4. The signals at 2922 $cm^{-1}$ and 2853 $cm^{-1}$ are due to the asymmetric and symmetric $CH_2$ stretching modes. The peak at 3008 $cm^{-1}$ is attributed to a cis HC=CH arrangement, while at 3076 $cm^{-1}$ is a trans arrangement [6]. Another proof of the organic capping layer are the signals at 1467, 1370 and 720 $cm^{-1}$, characteristic of a long organic chain, and at 998 and 908 $cm^{-1}$, due to $CH_2$ bending modes. 1526 $cm^{-1}$ and 1407 $cm^{-1}$ are attributed at the carboxylate group, and this confirms the presence of oleic acid. The apparition, after 30 min, of the new signals at 1710 $cm^{-1}$, the stretching of a carbonyl group (C=O), and 1646 $cm^{-1}$ suggest the presence of an amide [7]. The only way to have it, is a reaction between the oleylamine and the oleic acid. In fact, this signal is not present in the FT-IR spectra of $Bi_2S_3$-HMS and $Bi_2S_3$-NW samples, synthesized in presence of only OA and OAm, respectively. The high steric hindrance of the chains, can cause a multi-layer capping with some functional group facing the apolar solvent, decreasing the stability of the particles.

## 3. Conclusions

In summary, we presented different synthesis methods to produce $Bi_2S_3$ in orthorhombic phase with different size and morphologies. The $Bi_2S_3$-HMS method is the standard one and most employed for the production of photovoltaic devices. It allows a variation in size of the particles and so a control on the quantum size effect of the material. The other methods employ less aggressive chemicals and allow to change the morphology going from nanorods to nanowires, a fact that could prove useful to tune the conductive and optical properties. We have investigated the effect of the oleylamine on the growth of the particles and on its stability. Our results may prove useful to develop a dependable and "green" synthesis route for the production of colloidal $Bi_2S_3$ nanocrystals.

## Acknowledgments

R. P. gratefully acknowledges Sardinia Regional Government for the financial support of her PhD scholarship (P.O.R. Sardegna F.S.E. Operational Programme of the Autonomous Region of Sardinia, European Social Fund 2007-2013.

## References

[1] Aresti M et al *Colloidal $Bi_2S_3$ nanocrystals: quantum size effects and midgap states* 2014, Adv. Func. Mat. **24**, pp 3341-3350
[2] Konstantatos G et al, 2008, Nano Lett., **8**, 4002
[3] Hongbo L, 2010, CrystEngComm, **12**, pp 1127-1133
[4] Ibañez M et al, 2011, J. Phys. Chem. C, **115**, pp 7947-7955
[5] Cademartiri L et al, 2008, Angew. Chem. Int. Ed. Engl., **47**, 3814.
[6] Cannas C et al, 2010, Chem. Mater., **22**, pp 3353-3361
[7] John C, 2000, *Interpretation of infrared spectra, a practical approach*, Encyclopedia of Analytical Chemistry R.A. Meyers (Ed.) pp 10815-10837

Journal of Physics: Conference Series **566** (2014) 012018 doi:10.1088/1742-6596/566/1/012018

# Influence of phosphorous precursors on spectroscopic properties of $Er^{3+}$-activated $SiO_2$-$HfO_2$-$P_2O_5$ planar waveguides

**I Vasilchenko[1,2], A Carpentiero[1], A Chiappini[1], A Chiasera[1], A Vaccari[3], A Łukowiak[4], G C Righini[5,6], V Vereshagin[7] and M Ferrari[1,6]**

[1]IFN – CNR CSMFO Lab., Via alla Cascata 56/C Povo, 38123 Trento, Italy
[2]Dipartimento di Fisica, Università di Trento, via Sommarive 14, Povo, 38123 Trento, Italy
[3]FBK - ARES unit, Via alla Cascata, 56/C Povo, 38123 Trento, Italy
[4]Institute of Low Temperature and Structure Research, PAS, Okolna 2, 50-422 Wroclaw, Poland
[5]IFAC - CNR, MiPLab., 50019 Sesto Fiorentino, Italy
[6]Museo Storico della Fisica e Centro di Studi e Ricerche "Enrico Fermi", Piazza del Viminale 1, 00184 Roma, Italy
[7]Institute of High Technology Physics, National Research Tomsk Polytechnic University, 30 Lenin Avenue, 634050, Tomsk, Russia

E-mail: iustyna.vasilchenko@unitn.it

**Abstract.** $(70-x)SiO_2$-$30HfO_2$ -$xP_2O_5$ (x= 5, 10 mol %) glass planar waveguides activated by 0.5 mol% $Er^{3+}$ ions were prepared by sol–gel route. Several phosphorous precursors have been investigated for the synthesis of a dielectric stable sol useful for the realization of planar waveguides. The waveguides were investigated by different diagnostic techniques. The optical properties such as refractive index, thickness, number of propagating modes and attenuation coefficient were measured at 632.8 and 543.5 nm by prism coupling technique. Transmission measurements were carried out in order to assess the transparency of the deposited films. Photoluminescence measurements and lifetime decay curves of the $Er^{3+}$ transition ($^4I_{13/2} \rightarrow {}^4I_{15/2}$) were performed in order to investigate the role of $P_2O_5$.

## 1. Introduction

Silica glass ($SiO_2$ glass) is one of the most important materials in photonics [1]. Considering that it has many attractive properties, such as high UV transparency, high mechanical strength, high glass transition temperature, extremely low thermal expansion, and low reflective index, it has various potential uses in many optical devices, such as in waveguides and lasers. However the needs for compact and efficient photonic devices obtained by rare-earth-activated-glasses still drive the research of novel glass composition and optimized fabrication protocols [2, 3]. Considering the case of silica glass-based Erbium Doped Waveguide Amplifiers (EDWAs) their short length (around few centimetres) generally imposes a high $Er^{3+}$ doping level which may produce clustering effects [4]. These phenomena, due to the ion-ion interactions, lead to undesirable luminescence quenching, thus reducing the performance of the amplifier. At the same time the difficulty of low solubility of rare-earth in silica is due to the mismatch in size and valence between the rare-earth ions and the constituents of the silica network. A possible way to increase the amounts of rare-earth ions in the matrix avoiding or reducing clustering effects is the addition of co-doping agents, such as $P_2O_5$ or $Al_2O_3$ [5, 6]; their role resides in non-bridging oxygen's formation that benefits better rare-earth ions incorporation. Moreover from a technological point of view physical and chemical deposition techniques such Flame Hydrolysis Deposition (FHD) and Plasma Enhanced Chemical Vapor Deposition (PECVD) have important drawbacks associated to the incorporation of rare earth ions in silicate glasses due to the high process temperatures in the first case [7], and the low vapour pressure of most erbium compounds in the second [8]. Therefore there is a need to investigate complementary approaches for depositing Er-doped silicate glass, which can incorporate the full range of codopants needed to tailor the Er environment. One possibility is the sol-gel route, which allows very flexible chemistry and low process temperatures.

In this paper we have investigated planar waveguides with compositions (70-x) $SiO_2$-30$HfO_2$ -x$P_2O_5$ (x= 5, 10 mol %), activated by 0.5 mol% $Er^{3+}$ ions. We used different phosphorus precursors focusing the attention of their influence on the optical and spectroscopic properties.

## 2. Experimental setup

Multilayer wave guiding films activated with rare-earths were produced by sol-gel approach employing the motor driven dip-coating set-up. For film-forming solution (70-x)$SiO_2$-30$HfO_2$ -x$P_2O_5$ (x= 5, 10 mol %), activated by 0.5 mol% $Er^{3+}$ ions was prepared by mixing tetraethylorthosilicate (TEOS), ethanol, deionised water and hydrochloric acid as a catalyst, then pre-hydrolysed for 1 hour at 65 °C. A second solution was obtained by dissolving $HfOCl_2{\bullet}8H_2O$ as hafnia precursor in ethanol and then added to the TEOS solutions. Phosphorous and erbium $Er(NO_3)_3{\cdot}5H_2O$ precursors were mixed with final solution. Synthesis occurred under continuous stirring for 16 h at room temperature. The films were deposited on clean silica substrates with roughness of about 2 nm. Films were formed by dip-coating, with a dipping rate of 40 mm/min. After each dip, layer was subjected to annealing in air for 50 s at 900 °C. After a 10-dip cycle, the film was heated for 2 min at 900 °C after each tenth layer. In order to obtain a fully densified structure a further annealing at 900 °C for 5 min was performed [2]. The thickness and the refractive index at 632.8 and 543.5 nm were measured in TE polarization by an m-line apparatus based on the prism coupling technique [9, 10]. The losses at 632.8 nm were evaluated by photometric detection of the light intensity scattered out of the waveguide plane exciting the transverse electric $TE_0$ mode. Photoluminescence spectroscopy was performed using the 514.5 nm line of an Ar+ ion laser as excitation source. The luminescence was dispersed by a 320 mm single-grating monochromator with a resolution of 2 nm. The light was detected using a Photo Multiplier Tube (PMT) and standard lock-in technique [11]. Decay curves were obtained recording the signal by a digital oscilloscope.

## 3. Results and discussion

The choice of a suitable phosphorous precursor (PP) is a key point towards the realization of low losses waveguides operating at 1.5 μm with high $Er^{3+}$ ions content. Several PPs (reported in table 1) have been investigated for the synthesis of a dielectric stable sol useful for the realization of planar waveguides.

The composition and the optical parameters of the $Er^{3+}$-activated silica-hafnia-phosphorous oxide planar waveguides (called Wx) are reported in table 1.

**Table 1.** Phosphorous precursors and optical properties of $SiO_2$–$HfO_2$–$P_2O_5$ planar waveguides activated with 0.5% mol of $Er^{3+}$ ions.

| Label | Phosphorous precursors | $P_2O_5$ mol % | Layers number | Thickness ±0.3 μm | | Refractive index ±0.0005 | | TE Modes number |
|---|---|---|---|---|---|---|---|---|
| | | | | @632.8 nm | @543.5 nm | @632.8 nm | @543.5 nm | |
| W1 | $C_6H_{15}O_3P$ | 5 | 5 | - | 0.2 | - | 1.5762 | 1 |
| W2 | $(CH_3CH_2O)_3PO$ | 10 | 18 | 0.8 | 0.8 | 1.5868 | 1.5940 | 2 |
| W3 | $(CH_4CH_9O)_3PO$ | 5 | 18 | 0.9 | 0.9 | 1.5716 | 1.5871 | 2 |
| W4 | $H_3PO_4$ | 5 | 30 | 1.4 | 1.4 | 1.5585 | 1.5676 | 3 |

The systems W1, W2 and W3 present an opaque and whitish appearance attributable to the low solubility in the matrix of the PPs, indicating that these reagents are not suitable for the realization of optical devices.

The use of $H_3PO_4$ as phosphorous precursor permits to realize a waveguide (W4) that present a good transparency: over 92% in a wide range from 400 to 2000 nm, as evidenced in figure 1 by means of transmission measurement and allows to realize thick guiding film that support single mode at 1.5μm.

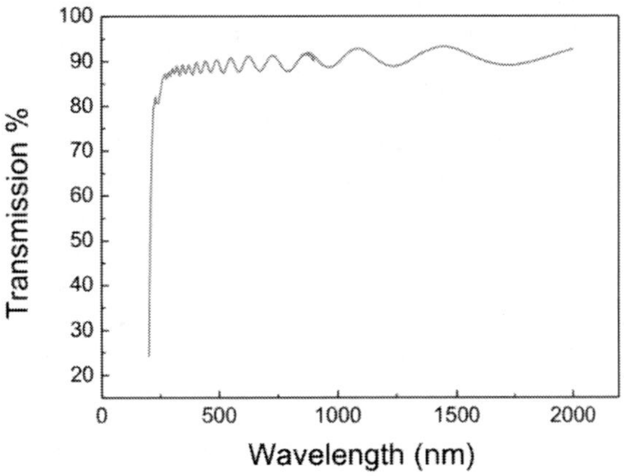


**Figure 1.** Transmission spectrum of the planar waveguide W4 doped by 0.5 mol % $Er^{3+}$ ions.

Figure 2 and figure 3 report the profile of the refractive index and of the $TE_0$ mode squared electric field of the waveguide W4, respectively. The $TE_0$ mode is computed for 632.8 nm and 1542 nm for the choice of the parameters determined by the m-line measurement.

The refractive index profile of the W4 waveguide can be reconstructed from the effective refractive indices at 632.8 nm by an inverse Wentzel–Kramers–Brillouin method [12], in particular from figure 2 we can observe that the W4 exhibits a single step profile with a uniform refractive index throughout the whole thickness.

The modelling of the square electric field, reported in figure 3, indicates that the optical parameters of the waveguide, i.e. refractive index and thickness, appear appropriate for application in the third telecommunication window (λ=1.5 μm). In fact, the confinement coefficient of the $TE_0$ mode, defined as the ratio of the intensity in the waveguiding film to the total intensity, which includes also the squared evanescent field, is 0.98 and 0.87 at 632.8 nm and 1542 nm, respectively.

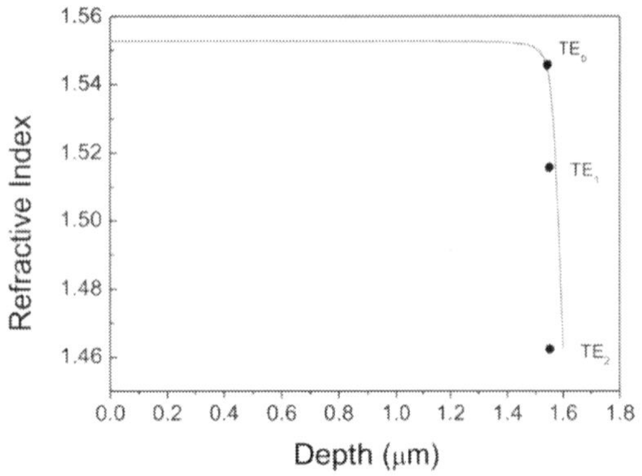


**Figure 2.** Refractive index profile of the W4 waveguide reconstructed from modal measurements at 632.8 nm for the TE. The effective indices of the TE (•) modes are reported.

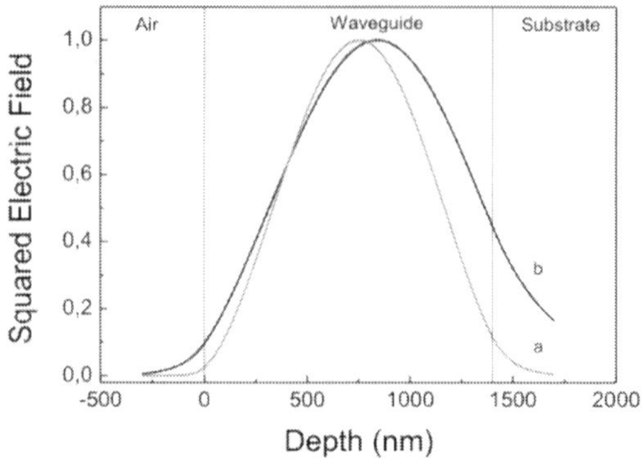


**Figure 3.** Calculated squared electric field profiles of the $TE_0$ mode at 632.8 and 1542 nm (a) and (b), respectively, across the layered structure, cladding of air, waveguide and the $SiO_2$ substrate of the W4 planar waveguide.

Photoluminescence measurements were performed in order to investigate the effect of $P_2O_5$, comparing the spectroscopic features of the W4 with those of a waveguide with the following molar composition $70SiO_2$-$30HfO_2$ activated with 0.5% mol (WU), doped with the same $Er^{3+}$ ions concentration.

In figure 4 is reported the photoluminescence (PL) spectra of transition ${}^4I_{13/2} \rightarrow {}^4I_{15/2}$ of $Er^{3+}$ ions acquired from the sample WU and W4. Analyzing figure 4 is possible to observe that luminescence shape is similar for both the samples with a main emission peak at 1533 nm and a spectral bandwidth of 48 nm, measured at 3 dB from the maximum of the intensity. The shape of the emission spectrum is characteristic of the ${}^4I_{13/2} \rightarrow {}^4I_{15/2}$ transition of $Er^{3+}$ ions in silicate glasses.

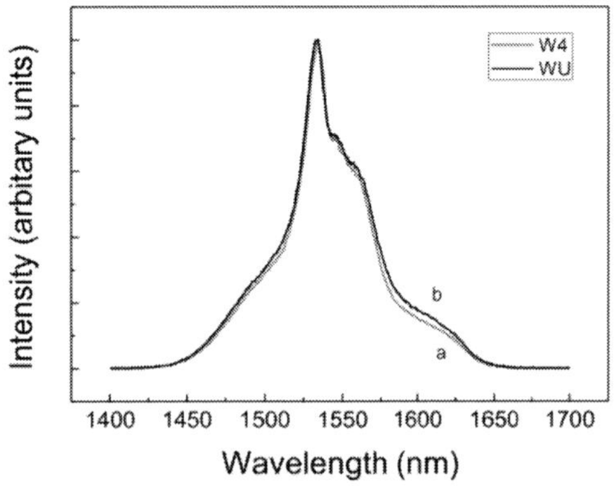


**Figure 4.** Room temperature PL spectra relative to the ${}^4I_{13/2} \rightarrow {}^4I_{15/2}$ transition of $Er^{3+}$ ions for W4 (a) and WU (b) waveguides upon excitation at 514nm.

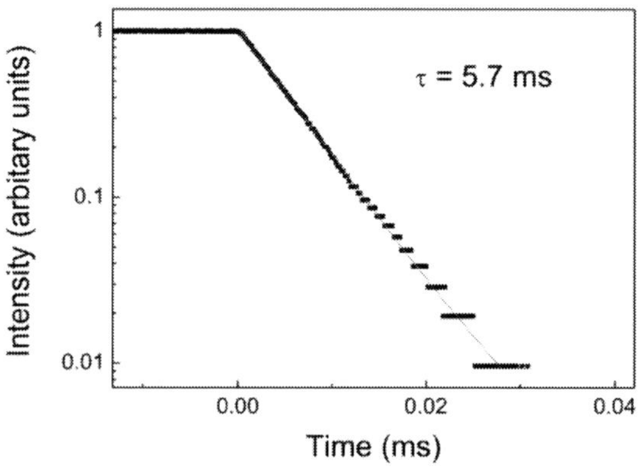


**Figure 5.** Decay curve of the luminescence from the ${}^4I_{13/2} \rightarrow {}^4I_{15/2}$ metastable state of $Er^{3+}$ ions for the W4 waveguide. The red solid line represents single exponential fits to the decay data.

Figure 5 reports the luminescence decay curve from the metastable level ${}^4I_{13/2} \rightarrow {}^4I_{15/2}$ for the W4 waveguide obtained upon excitation at 514 nm; we can notice that the decay curve exhibits a single exponential behavior, indicating that $Er^{3+}$ ions all occupy sites characterized by similar local environment and the value of the measured lifetime $\tau_{W4}$= 5.7 ± 0.1 ms. This value can be compared with that obtained on the sample WU ($\tau_{Wu}$= 5.1 ± 0.1 ms) and discussed in ref [13]. The increase of the lifetime measured for the sample W4 with respect to WU can be attributed to the presence of $P_2O_5$ in the matrix indicating that the phosphorous oxide can increase the $Er^{3+}$ ions solubility.

## 4. Conclusions

In conclusion $Er^{3+}$ activated $65SiO_2$-$30HfO_2$-$5P_2O_5$ waveguides (WGs) were prepared via the sol–gel method and dip coating processing. We have studied the influence of the phosphorous precursors on the optical properties of planar waveguides demonstrating that $H_3PO_4$ precursor allows to obtain a guiding transparent film. The waveguide realized (W4) supports a single and well confined (87%) mode at 1.5μm, presents a single step refractive index profile and an attenuation coefficient of about 3.5dB/cm at 632.8 nm.
Luminescence in the third telecom region, with a spectral width of 48 nm and a single exponential lifetime of 5.7 ms, was observed upon excitation at 514.5 nm.
Finally the increase of the lifetime of the W4 with respect to that obtained on a complementary WG not codoped with $P_2O_5$ indicates that the phosphorous oxide can increase the $Er^{3+}$ ions solubility reducing the concentration quenching.

## 5. Acknowledgments

This research is performed in the framework of MAE "Smart optical nanostructures for green photonics" (2013–2015) and CNR-PAS (2014-2016) projects.

## References

[1] Righini G C, Pelli S, Ferrari M, Armellini C, Zampedri L, Tosello C, Ronchin S, Rolli R, Moser E, Montagna M, Chiasera A, and Ribeiro S J L 2002 *Optical and Quantum Electronics* 34 1151

[2] Righini G C and Ferrari M 2005 *Riv. Nuovo Cimento* **28** 1

[3] Righini G C and Chiappini A 2014 *Opt. Eng.* **53** 071819

[4] Righini G C, Berneschi S, Nunzi Conti G, Pelli S, Moser E, Retoux R, Féron P, Gonçalves R R, Speranza G, Jestin Y, Ferrari M, Chiasera A, Chiappini A, and Armellini C 2009 *J. Non-Cryst. Solids* **355** 1853

[5] Deschamps T, Ollier N, Vezin H, and Gonnet C 2012 *J. Chem. Phys.* **136** 0145031

[6] Lemaire P J, MacChesney J B and Simpson J R 1989 *US4830463 A* EP0328315A1

[7] Kik P G and Polman A 1998 *MRS Bull.* **23** 48

[8] Ribeiro S J L, Messaddeq Y, Gonçalves R R, Ferrari M, Montagna M, Aegerter M A 2000 *Applied Physics Letters* **77** 3502.

[9] Tien P K and Ulrich R 1970 *JOSA* **60** 10

[10] Shuto K, Hattori K, Kitagawa T, Ohmori Y, and Horiguchi M 1993 *Electron. Lett.* **29** 139

[11] Gonçalves R R, Carturan G, Montagna M, Ferrari M, Zampedri L, Pelli S, Righini G C, Ribeiro S J L, Messadeq Y 2004 *Opt. Mat.* **25** 131

[12] Chiang K S 1985 *J. Lightwave Technol.* **LT3** 385

[13] Chiappini A, Chiasera A, Armellini C, Varas S, Carpentiero A, Mazzola M, Moser E, Berneschi S, Righini G C, Ferrari M 2011 *J. Sol-Gel Sci. Technol.* **60** 408

Journal of Physics: Conference Series **566** (2014) 012019 doi:10.1088/1742-6596/566/1/012019

# Use of GPUs to boost the performance of a lattice-free tumour growth model

**Sabrina Stella[1], Roberto Chignola [2] and Edoardo Milotti[1,3]**

[1]Department of Physics, University of Trieste, Italy
[2]Department of Biotechnology, University of Verona, Italy
[3]Istituto Nazionale di Fisica Nucleare, Sezione di Trieste, Italy

E-mail: Sabrina.Stella@ts.infn.it

**Abstract.** We recently developed a computational model of tumour growth. It is a cell-based model that can simulate the growth of multicellular tumour spheroids up to more than one million cells. The simulation program is very demanding and simulation time severely limits the integration of additional biological details, and indeed, at the moment, a typical simulation run requires tens of days to be completed. A new version of the code that exploits Graphics Processing Units (GPUs) to boost performance is being developed. In this paper we describe the design and implementation of a nearest-neighbour search (NNS) algorithm suitable to run on GPU. The algorithm will be integrated in the original code to manage the geometrical calculation in the simulation of the spheroid. Initially the stand alone NNS algorithm was tested for spheroids of different size: better efficency was obtained for bigger spheroids. Eventually the code was integrated in the whole simulation code and preliminary runs gave a speed up of about 5 for spheroids of relatively small size (15000 cells).

## 1. Introduction

Many computational problems in science and technology require the use of high performance computing solutions, since the huge amount of data and the complexity of the problem cannot be managed with single machines. Researchers then turn to highly parallel super-computer clusters, available in High Performance Computing (HPC) centres to solve their computational problems. A few years ago a new solution came from the video-game industry with the construction of small highly parallel processors, the Graphics Processing Unit (GPU). These are graphic accelerators designed to render scenes – that is to generate images – and relieve the computational load of a CPU, as a consequence more accurate images can be achieved in shorter time. The use of GPUs for scientific computing was not immediately possible, because these devices were designed to solve only those simple and repetitive tasks that are characteristic of the rendering process. Nevertheless, thanks to pioneering efforts and to the increasing interest of the scientific community, vendors were encouraged to propose new GPU architectures and programming models, to utilise these devices for more general problems. In 2007 the advent of the Nvidia CUDA architecture brought a considerable change in this scenario, with a new generation of graphic cards that could be easily programmed with extensions to standard C++ or FORTRAN code [1]. Nowadays, the community of CUDA programmers is steadily growing, just as the interest in the design of parallel algorithm suited to the GPU architecture [2, 3]. Not every algorithm can benefit from the computational power of the GPUs, because even those

Journal of Physics: Conference Series **566** (2014) 012019 doi:10.1088/1742-6596/566/1/012019

algorithms that can be parallelised are not always portable on the GPU architecture. GPUs were originally designed for the so-called data-intense parallel problem, such as the rendering processes, where the same flux of instructions is executed on many cores on a huge amount of data. For this reason, even today, graphic cards are more suitable to run those algorithms that can be *data-parallelised* or that can be subdivided in small sub-problem based on group of data – and this happens to be the case of most computational problems in physical simulations.
The design of a proper parallel algorithm for GPU requires a basic knowledge of the device's architecture. The main difference between GPUs and CPUs is that the processing units in a GPU are simpler and smaller when compared to a traditional CPU, and this allows the integration of thousands of computing units in a GPU card, hence its computational power. In a recent GPU architecture called Kepler, the CUDA cores are grouped in multiprocessors (SMX) which have 192 cores (Tesla K20) and a single card can have more than one SMX (there are 13 SMXs in the Tesla K20). The GPU card has its own private memory with a high bandwidth. The memory is organised in a hierarchical structure: typically there is a high capacity and high latency access memory, called *global memory*, it is accessible by all the cores. Other different levels of memory are also present, they are characterised by lower latency time but also by lower capacity compared to the global memory and some of those are gradually more private to a single core (for details see [1]). The memory management task is partially left to the programmer who can obtain a better performance with a proper memory use.
A GPU program is organised in *threads*, each thread executes the same set of instructions – called a *kernel* – although on different data. Threads are grouped in *blocks*, and blocks are organized in a *grid* that can be 1-, 2- or 3-dimensional. The choice of the number of threads and the dimension of the grid and of the blocks – up to a hardware-dependent maximum – is left to the programmer. All this contributes to make the design and implementation of parallel algorithms for GPUs both interesting and useful, although not trivial. At present, parallel algorithms running on GPU are becoming quite widespread, both in research and in industry. In this paper we explain how we use the GPU to speed up the biophysical simulation of a numerical model of tumour growth.

## 2. Biophysical model of tumour growth

We have developed a biophysical model that simulates the growth of small avascular solid tumours. Originally, it was conceived to describe the evolution of a multicellular tumour spheroids (MCTS). A MCTS is an *in vitro* cell culture where cells are grown in such a way to prevent their adhesion to the walls of the culture well, and they cluster in a roughly spherical shape. Our model starts with the mathematical description of the individual cell cycle. Each cell is approximated as a soft sphere that receives nutrients from the environment, it can grow and proliferate and it interacts, both biochemically and mechanically, with the surrounding cells. The model includes a set of reaction-diffusion equations that describe the dynamics of a selected set of molecules. Threshold mechanisms are used to mark the transitions between the phases of the cell cycle, and a phenomenological model describes the cell-cell forces, which exhibit a repulsive or attractive behaviour according to the distance between the cells' centers. For a more complete description of the model see [4]. In a standard run we start with a single cell in a uniform environment, and as the run proceeds, cells grow and proliferate and eventually the cell aggregate can include as many as 1.5 million cells. The results of simulation runs show that an *in-silico* MCTS has properties comparable with those of real MCTS, such as its external shape and the distribution of dead cell in the aggregate (presence of a necrotic core) [5].

The program that implements the model is coded in C++, the disordered structure is managed with the CGAL library [6], its computational complexity depends roughly linearly on the number of cells. However, the number of cells grows fast, the initial growth is exponential in time, and it turns out that the CPU time for each simulation step steadily increases as

the simulated spheroid grows, and that a full simulation run takes tens of days. During the simulation, the spheroids do not evolve following a predetermined lattice structure but the cells are free to occupy every position on the 3D space. As a result, the computational complexity of each step depends in good part on the geometrical calculation performed with CGAL. As a consequence, the simulation is more realistic but requires the continuous evaluation of some geometrical parameters which are needed to solve the model's equations. In particular the list of neighbouring cells - of each cell in the spheroid - needs to be evaluated whenever the position of the cell's center changes beyond a given threshold; the movements are due to cell-cell forces and to cell division, when the spheroid is large these movements are so frequent that the neighborhood changes at each time step.
To improve performance we modify the code by moving the geometrical calculation onto a GPU. This was done by developing and implementing a proper parallel algorithm for searching the nearest neighbour cells, starting with a set of cells that changes dynamically.

## 3. Design and Implementation of a GPU-based adjacent-cell search algorithm

The nearest-neighbour search is a well known problem in the field of computational geometry. It can be solved using two different approaches: the brute force approach – i.e., by comparing the distance between each single cell and all the other cells, and finally selecting those with the smallest distances – or by performing a *space partitioning operation* to facilitate the search – i.e., by limiting the search operation on a given subspace.
Parallel formulations well-suited to GPUs of both algorithms have already been developed [7, 8, 9] and most of the experience has been garnered in the field of image processing.
In our case we started with the algorithm described in [9] , which considers a space partitioning operation using a cubic grid and we modified it in order to find the "adjacent cells", i.e., those cells that are in close contact with a query cell, and therefore contribute to the diffusion processes. The contact between two cells exists only if the distance between their center is smaller than the sum of their radii multiplied by a suitable extension factor; this last parameter is introduced to take into account the softness of cells. The extended radius – i.e., the radius multiplied by the extension factor – is also used to evaluate the contact surface.

### *3.1. Adjacent cell search algorithm scheme*

The search of the adjacent cells in the whole spheroid can be easily parallelised on GPUs when threads are tasked to find the neighbourhood of individual cells. These operations are indeed independent of each other, therefore they can be executed in parallel. As already mentioned, the search operation is optimised by the space partitioning operation: first the bounding box of the spheroid is identified and then it is subdivided in equally sized small boxes (see fig.1). Next, each cell can be given a hash value, that is a number identifying the small box where it resides. This hash value can be used by the thread to identifyin subspace where performing the search operation, in particular each thread verifies the adjacency condition for the subset of cells included in the boxes having the same or adjacent hash values as the examined cell, that is it scans 27 small boxes (or less in case of cells belonging to the boundary boxes). A more complete description of the algorithm is given in [10].

### *3.2. Test run*

The algorithm stand alone was implemented using the extention CUDA 5.0 for C++ and the Thrust library [11] - the parallel version of the Standard Template Library. The program was run on a GPU Nvidia Quadro 4000, provided with 256 cuda core of about 0.95 GHz, (32 cuda core/SM). The execution is performed opening a number of threads, equal or higher than the number of cells composing the spheroid, i.e. a thread for each biological cell. The threads were grouped in block of 64 (a block multiple of 32 thread can assure better performance) and

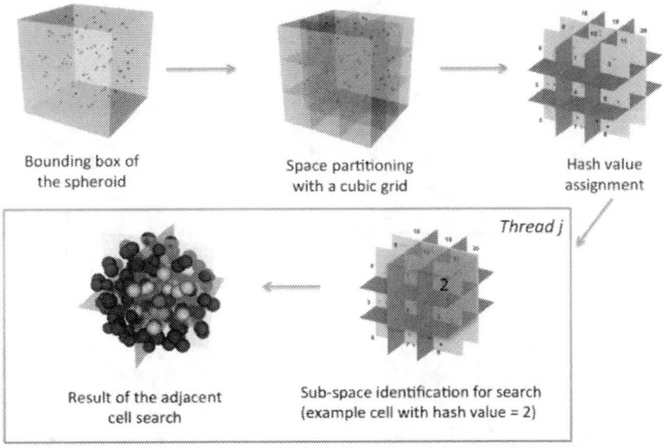


**Figure 1.** Main steps of the nearest-neighbour search (NNS) algorithm.

the blocks were arranged on a squared 2D grid in order to manage approximately $10^6$ cells or threads; the grid's dimension changes according to the size of the problem, as an example for a spheroid of 40000 cells we used a 25x25 grid of threads blocks. The algorithm was tested for spheroids with different size, from hundreds to one million of cells. As expected the efficency of the algorithm increases with the size of spheroid. The execution time (in ms, calculated by the CUDA runtime API `cudaEventElapsedTime()[12]`) for the kernel responsible for the adjacent cell search (the last two steps in fig.1) is shown in fig.2. The search operation requires only 403 ms to determine the list of neighbors for a spheroid higher than one million of cells.

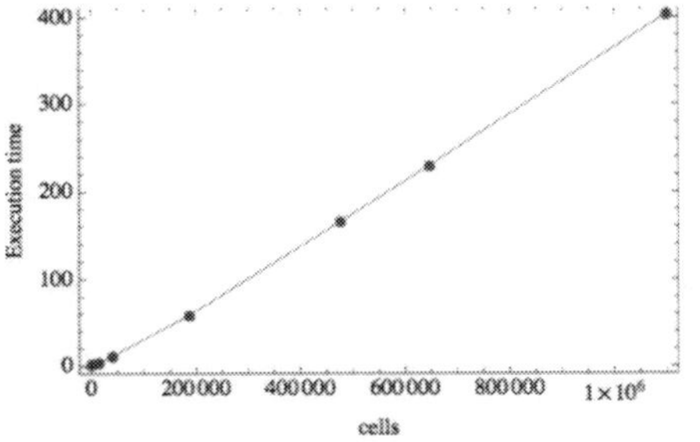


**Figure 2.** Execution time (ms) of the kernel responsible for the adjacent cell search as a function of the total number of cells (it does not include the space partitioning operation and the hash value assignment operation)

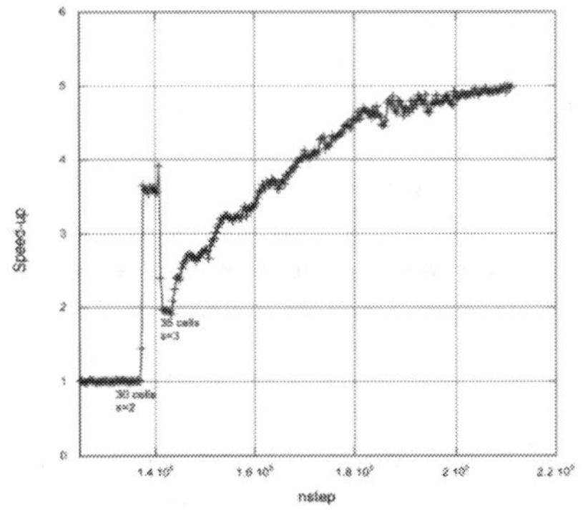


**Figure 3.** Speed up as a function of spheroid size for the geometrical calculation in the main program

*3.3. Accelerating the tumour growth runs on a GPU using the NNS algorithm*

A standalone version of the algorithm for the adjacent cell search was integrated in the simulation program. The main program is essentially a time loop which starts after initialisation and it performs in sequence the following operations: 1. calculation of biochemistry and biomechanics of the cluster; 2 check for discrete cellular events; and, finally, 3. recalculation of the geometrical parameters, including the proximity relationships, if a mitosis event occurs, or if the cluster changed shape beyond a certain threshold. The new CPU-GPU version of the code was implemented by partially assigning to GPU the third step of the main loop. This is done by transferring on GPU all the geometrical data for the cells aggregate (positions of cell centres and radii), and running the nearest-neighbour search (NNS) algorithm to find the list of the adjacent cells. Moreover, a new condition was introduced to determine also the list of boundary cells [1]. These cells represent the interface with the environment and their correct evaluation is important for a correct growth of the spheroids. After the calculation, the GPU transfers to the CPU two vectors: `AdCells[N*30]` and `BC[N]` – where `N` is the total number of cells and 30 is the maximum number of neighbours for each cells – they store respectively the index of the adjacent cells and the list of boundary cells. These device-host (GPU-CPU) and host-device transfer operations are usually the most expensive ones, and have to be executed sparingly to get the best performance.

To evaluate the performance of this new version of the code we introduce a speed-up value, that is the ratio of the execution time on CPU and on CPU+GPU of the functions responsible for the geometry calculation. A preliminary result is given in fig.3: the speed-up increases with the simulated time step or similarly with the size of the spheroids, and this is due to the higher efficiency of the nearest neighbour search algorithm for larger spheroids. In these preliminary runs simulations stop at a relatively small size (about 15000 cells), where the maximum speed-up obtained is about 5. We are now developing a new version of the algorithm for the boundary cells identification: this is required to perform a full simulation.

## 4. Conclusions

The preliminary runs of the new code show an increasing speed-up value as a function of the spheroid size. A better performance is expected when more recent GPUs, such as TeslaK20 will be used. The forthcoming introduction of some common optimisation procedures – such as a better use of the memory hierarchy – is also expected to increase the speed-up.We also expect a large gain by moving on GPU the other critical step, i.e., the reaction-diffusion part of the model. A faster code is fundamental for the introduction of new biological detail in our model, with the final goal to get a more realistic biological environment that may be used to reveal some hidden properties of the tumour growth process.

## References

[1] `http://docs.nvidia.com/cuda/cuda-c-programming-guide/`
[2] Couturier R 2014 *Designing Scientific Applications on GPUs* (CRC Press)
[3] Navarro C A, Hitschfeld N and Mateu L 2014 *Communications in Computational Physics* **15**
[4] Milotti E, Del Fabbro A and Chignola R 2009 *Computer Physics Communications* **180** 2166–2174
[5] Milotti E and Chignola R 2010*PLoS One* **5**
[6] `http://www.cgal.org`
[7] Garcia V, Debreuve E and Barlaud M 2008 *Computer Vision and Pattern Recognition Workshops, 2008. CVPRW'08. IEEE Computer Society Conference on* (IEEE) pp 1–6
[8] Pan J and Manocha D 2011 *Proceedings of the 19th ACM SIGSPATIAL International Conference on Advances in Geographic Information Systems* (ACM) pp 211–220
[9] Pedro L, Teixeira J M, Farias T, Reis B, Teichrieb V and Kelner J 2012 *International Journal of Parallel Programming* **40** 313–330
[10] Stella S, Chignola R, Dogo F and Milotti E 2014 *High Performance Computing and Simulation (HPCS), 2014 International Conference on* 621 – 626
[11] https://code.google.com/p/thrust/
[12] http://docs.nvidia.com/cuda/cuda-runtime-api

[1] The condition for a cell to be considered on the boudary is to lie on the outermost small box layer

Journal of Physics: Conference Series **566** (2014) 012020 doi:10.1088/1742-6596/566/1/012020

# An innovative radio-guided surgery technique for complete resection of tumors

**A Russomando [1,2,3], F Collamati [1,2], F Bellini[1,2], V Bocci[2], E De Lucia[4], R Faccini[1,2], P M Frallicciardi[2,5], M Marafini[2,5], I Mattei[2,6], S Morganti[2], V Patera[2,7], L Piersanti[2,7], D Pinci[2], L Recchia[2], M Senzacqua[1,2], A Sarti[4,7], A Sciubba[2,7], E Solfaroli Camillocci[3] and C Voena[1,2]**

[1]Dipartimento di Fisica, La Sapienza Università di Roma, Rome, Italy
[2] INFN Sezione di Roma, Rome, Italy
[3] Center for Life Nano Science@Sapienza, Istituto Italiano di Tecnologia, Rome, Italy
[4] Laboratori Nazionali di Frascati dell'INFN, Frascati, Italy
[5] Museo Storico della Fisica e Centro Studi e Ricerche E. Fermi, Rome, Italy
[6] Dipartimento di Fisica, Università RomaTre, Rome, Italy
[7]Dipartimento di Scienze di Base e Applicate per l'Ingegneria, La Sapienza Università di Roma, Rome, Italy

E-mail: andrea.russomando@roma1.infn.it

**Abstract.** Finding new ways to fight cancer is essential to increase the patients life expectancy. This paper reports the latest results of the project CHIRONE finalized to increase the potential of the Radio Guided Surgery through the use of $\beta^-$ emitting radio-tracers and $\beta^-$ probes. This innovation could overcome the present main limiting factor represented by a diffuse background due to the high penetration power of the gamma radiation used. We created a prototype of $\beta^-$ probe and in this paper we report measures of photon efficiency, acquired with commercial photons sources. Then we estimated the signal and background rates in realistic cases of meningioma through a simulation. The device is able to detect residuals of 0.1 ml in 1 s with an administered activity less than 3 MBq/kg.

## 1. Introduction

Cancer figures among the leading causes of death worldwide [1], so increasing the five-years relative survival rate could be a significant result from a clinical point of view. This rate is calculated as the percentage of patients with a disease that are alive five years after that their disease is diagnosed divided by the percentage of the general population of corresponding sex and age that are alive after five years.

In this contribution we propose an innovative Radio Guided Surgery (RGS) technique that allows to perform complete tumour resection, minimizing the amount of healthy tissue removed. The procedure is based on the administration of a radioactive tracer to the patient before the surgery [2]. A radioactive tracer is a chemical compound in which one or more atoms have been replaced by a radioisotope. The tracer is preferentially taken up by the tumour cells, increasing the radioactivity over the tumour masses with respect to the surrounding healthy tissues. This implies that each tumour requires its own radio-tracer.

After tumour removal, a real time scanning of the patient is performed by the surgeon, to check for residual radiation through specific probe systems.
A widely used marker for the tracer is the $^{99m}$Tc, a gamma emitter, that produces photons with an energy of 140 keV as decay products. Due to the body density, this energy implies a photon mean free path up till 10-15 cm severely limiting the RGS application to few specific sites (i.e. breast, melanoma) not significantly affected by tracer uptake background of nearby healthy organs. On the basis of these considerations, the use of different radio-marker tracers, like beta emitters, was proposed.
The choice of a beta minus emitter, like $^{90}$Y, would allow the emission of electrons with a reduced penetration power ($\sim$ cm, since their maximum energy is 2.2 MeV) in absence of gamma contamination. The physical characteristics of this particles are making $^{90}$Y particularly suitable for an application of the RGS to tumour families that are currently outside its field of application. A wide application of such novel technique will require the research of $\beta^-$ emitter radio tracers, since only few of them are currently well known and tested.

In this paper we report our progress in the development of different $\beta^-$ prototypes and their relative experimental performances. In particular we characterized the photon efficiency and we estimated the potential of the prototypes in clinical applications.

## 2. Methods

### *2.1. $\beta^-$ probe prototype*

A prototype of $\beta^-$ probe was developed to test the performance of RGS with $\beta^-$ radio-tracers. The prototype is described in detail in ref. [3]. The following targets were established:

- Real time data acquisition (essential during surgery)
- Detector high efficiency (to reduce the administered activity to the patient)
- Compact and handy device (easy of use in complex clinical situations)

An organic crystal, the para-terphenyl [4] doped with 0.1% diphenylbutadiene was chosen as detector. The low density (1.16 $g/cm^3$) of the material implies a scarce sensitivity to *Bremsstrahlung* emission. The detector has been tested with $^{60}$Co, $^{137}$Cs and $^{133}$Ba photon source, with a copper shield placed in front of the probe. The shield was required to stop the electrons that were produced by the sources, since they were not pure gamma emitters. The efficiency of the probe has been measured as the ratio between the photons' nominal activity and the measured counts, taking into account the presence of the shield.
Two probes, fit for operating on tumour areas of different dimensions, were realized to make possible the application in various clinical situations. The availability of alternative solutions could be helpful for the device compliance to strict medical rules. For localized tumours a para-terphenyl cylinder of 5.1 mm diameter and 3 mm height was used coupled with optical fibres that carried out the light to a photomultiplier tube (probe S4). For wider areas the probe was designed with a cylinder of 10 mm diameter and 3.2 mm height, directly coupled to a silicon photomultiplier (probe SiPM).

Portable electronics based on Arduino Due, with wireless connection to PC or tablet was used for the readout. The Arduino Due is a microcontroller board, based on a 32-bit ARM core microcontroller. The board can operate on an external supply of 6 to 20 volts (for an introduction about Arduino, see ref. [5]).

Ad-hoc phantoms [Figure 1] were conceived for the control experiments using sponge material bathed with a radioactive saline solution (mixture of distilled water and $^{90}$Y) in which the activity was modulated by the dilution. The phantoms were easy to manage and were assembled to create complex patterns to simulate different clinical situations.

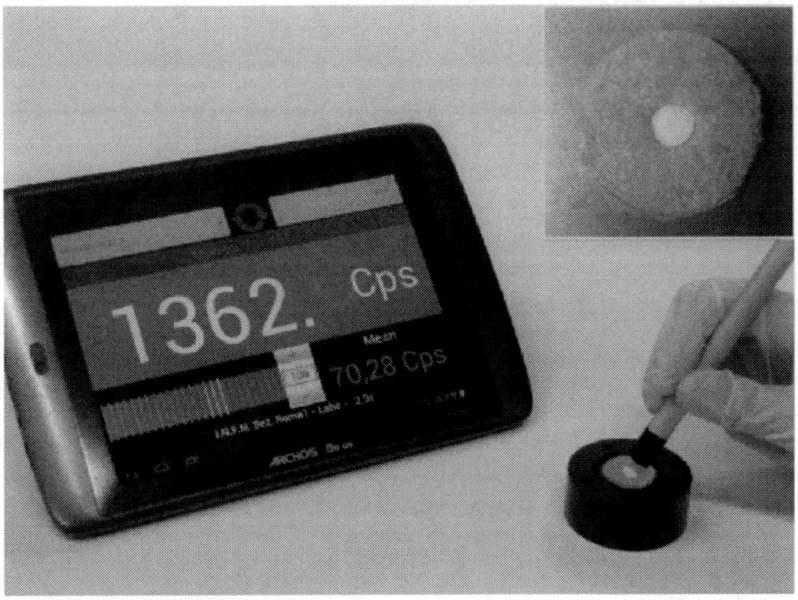


**Figure 1.** Operator using the probe over a phantom. The result of the data acquisition is displayed on a tablet. The detector is located inside the black cover at the top of the probe. Inset: phantom. The two different sections (yellow and blue) have different activities, to simulate the different uptakes between the tumour and the surrounding healthy tissue of the clinical case in exam. The yellow phantom has a 5 mm diameter and a 2.5 mm height. The blue one has the same height, but a 20 mm diameter.

*2.2. Monte Carlo simulation*

A Monte Carlo simulation, based on FLUKA, was used to investigate the probe performance for a real surgery application. FLUKA is a Monte Carlo package used to simulate interactions between particles and matter [6].

The smallest volume detectable by a PET scan corresponds to 0.1 ml. In our simulation the equivalent tumour mass was represented by a cylinder with a 3 mm radius and a 3.5 mm height. The residual was surrounded by an extended region with a lower uptake [Fig 2]. The uptakes of the radio-tracer for the tumour mass and the surrounding health tissues were calculated from PET images of 10 patients affected by meningioma brain tumour. DOTATOC, a synthetic somatostatin analogues, marked with $^{90}$Y was used as radio-tracer for this simulation. An activity of 3 MBq/Kg, the normally applied dose for PET scan, was used as benchmark. The simulations provided the rates on probe for the signal and the noise for each clinical case. Signal was defined as the rate over the tumour mass, while background as the rate over the healthy tissue.

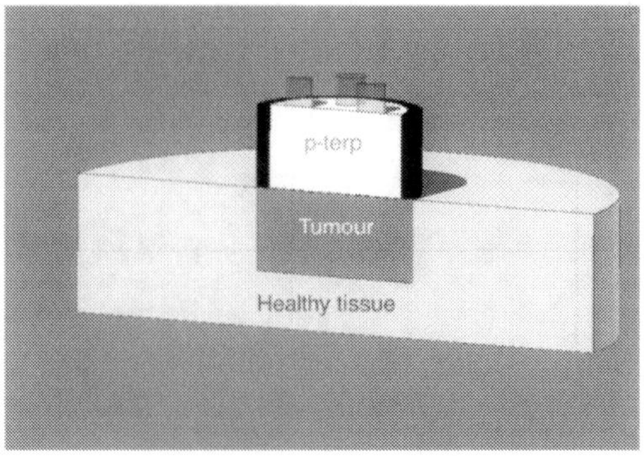


**Figure 2.** Simulation set-up. A tumour residual (red) surrounded by a healthy tissue (pink) with a lower uptake. The probe consists of a cylinder of para-terphenyl (white) surrounded by a PVC shield (black).

## 3. Results

### *3.1. Photon efficiency*

We used probe SiPM to estimate the photons efficiency, since it has the higher geometrical efficiency. We estimate an efficiency from $60.5 \cdot 10^{-6}$ to $6.1 \cdot 10^{-6}$ for photons from 1330 keV to 356 keV, with the efficiency that decreases according to the energy. Considering the probability of *bremsstrahlung* ($\sim 10^{-3}$) and its energy spectrum [Figure 3] we confirmed that the probe is insensitive to the photons.

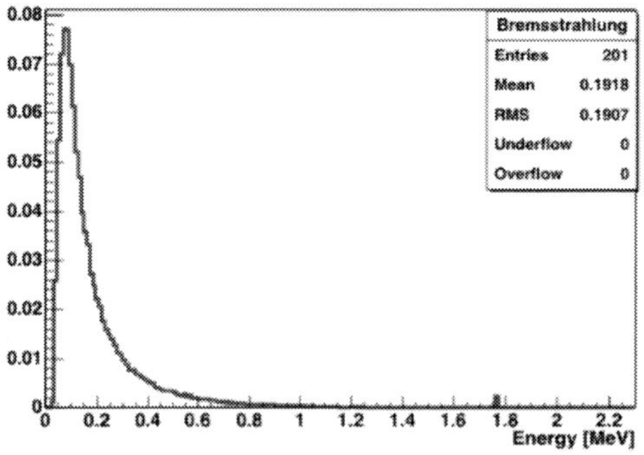


**Figure 3.** Simulated photons energy spectrum obtained by the simulation of a cylinder with the equivalent density of the human body filled with a diffuse source of $^{90}Y$. The photons are due to *bremsstrahlung* contribution

### *3.2. Estimated rates on probe S4*

Table 1 shows the simulations results for each patient of the rates on probe S4 estimated using the Monte Carlo simulation. Rates were converted in the minimum time necessary to diagnose the fraction of tissue under the probe with a false positive probability less than 1% (due to background fluctuation simulating a tumour mass) and a false negative probability less than 5% (probability to miss the residual). Taking into account the fluctuations due the poissonian nature of the process, the minimum time was established as the minimal value of time for which

**Table 1.** Results of the simulation. For each patient the table shows the probe rates for the signal (S) and for the background (B) for each lesion. The last column shows the minimum time (T) that the surgeon has wait to be sure to detect the residual. These results are calculated for an administered doses of 3 MBq/kg for meningioma tumour.

| Patient ID | S (Hz) | B (Hz) | T (s) |
|---|---|---|---|
| 1 | 32.2 | 1.9 | 0.2 |
| 2 | 17.6 | 2.6 | 0.6 |
| 3 | 33.7 | 3.5 | 0.3 |
| | 50.3 | 3.5 | 0.3 |
| | 76.8 | 3.5 | 0.1 |
| 4 | 89.4 | 4.5 | 0.1 |
| 5 | 66.7 | 4.4 | 0.2 |
| | 53.2 | 4.4 | 0.2 |
| | 57.6 | 4.4 | 0.2 |
| 6 | 107.6 | 1.8 | 0.1 |
| | 56.1 | 1.8 | 0.2 |
| 7 | 50.2 | 3.9 | 0.2 |
| 8 | 55.7 | 3.6 | 0.2 |
| | 31.2 | 3.6 | 0.2 |
| | 29.6 | 3.6 | 0.4 |
| 9 | 13.4 | 2.4 | 0.9 |
| | 15.1 | 2.4 | 0.7 |
| 10 | 14.6 | 1.2 | 0.6 |
| | 12.6 | 1.2 | 0.8 |

existed a value of rates threshold for which both the requests were satisfied. This time resulted to be lower than 1 s in all the cases. The administered dose was 3 MBq/kg.

## 4. Conclusion

In this contribution we presented the results obtained from a Monte Carlo simulation of a $\beta^-$ probe implementing a novel technique for RGS.

The simulation was based on the result of phantom test and showed that we would be able to identify the tumour mass of the volume of 0.1 ml in 1 s at 95% C.L. after the administration of an activity at below 3 MBq/Kg (for patient affected by meningioma), fully supporting the technique potential. An indication that the dose delivered to the patient could be reduced has been obtained and will be explored with dedicated R&D in the future.

## References

[1] Jemal A, Ma J, Siegel R and Zou Z 2014 Global cancer statistics *CA: a cancer journal for clinicians* **64** p 9
[2] Mariani G, Giuliano A E and William Strauss H 2007 Radio-guided surgery, a comprehensive team approach *Springer*
[3] Solfaroli Camillocci E et al. 2014 A novel radioguided surgery technique exploiting $\beta-$ decays *Scientific Reports* **4** 4401
[4] Angelone M et al. 2014 Properties of para-terphenyl as detector for alpha, beta and gamma radiation *IEEE Transactions on Nuclear Science* **61** p 1483
[5] Banzi M 2011 Getting Started with Arduino, Second Edition *OReilly Media*

[6] Ferrari A, Sala P R, Fasso A and Ranft J 2005 FLUKA: a multi particle transport code *SLAC-R*-773

Journal of Physics: Conference Series **566** (2014) 012021 doi:10.1088/1742-6596/566/1/012021

# Degradation of Cd-yellow Paints: ab initio study of the adsorption of oxygen and water on {10.0} CdS surface

**Laura Giacopetti**
CNR-IOM Cagliari and Dept. of Physics, University of Cagliari, Cagliari, Italy

**Alessandra Satta**
CNR-IOM Cagliari, c/o Dept. of Physics, University of Cagliari, Cagliari, Italy

E-mail: laura.giacopetti@dsf.unica.it

**Abstract.** The cadmium yellow paints (CdS) used in impressionist and modernist paintings in early 1900s are undergoing several deterioration processes, including whitening and discoloration. A relevant effect produced at the surface of the paintings is the growth of discolored crusts, formed mainly by white globular hydrated cadmium sulfate ($CdSO_4*nH_2O$) and cadmium carbonate ($CdCO_3$). Recent studies ascribe to an initial photo-oxidation process of CdS the input for the formation of such whitish compounds. In order to understand, at atomic level, the oxidation and hydration mechanisms of these whitish globules, we present the early stages of a theoretical study of the interaction between the hexagonal {10.0} surface of CdS and $O_2$ and $H_2O$ molecules to simulate the combined effects of exposure to air and humidity. For this purpose, we adopted a first principles method within the framework of the Density Functional Theory (DFT) in the Generalized Gradient Approximation (GGA-PBE) with the use of ultrasoft pseudopotentials.

## 1. Introduction

Among the inorganic pigments, cadmium sulfide has high tinting, covering power, wide applicability, high stability in oil and water, pure hues ranging from very light yellow to orange, red and deep maroon-red. The cadmium pigments are meant to be cadmium sulfide crystallized in a hexagonal wurtzite lattice [1]. Cadmium sulfide with yellow color and high tinting forms the basis for all cadmium pigments. Consequently, important 19th to 20th century artists such as Claude Monet, Vincent Van Gogh, and Pablo Picasso frequently employed CdS in their paintings. Even though the cadmium sulfides are generally regarded as stable, various cases of specific deterioration noted on paintings have indicated that not all varieties are permanent. The deterioration occurs primarily in the lighter yellows, and involves discoloration (fading or whitening and a darkening) and loss of adhesion within the paint itself, producing chalking and crumbly surfaces. Phenomenons of deterioration are most pronounced in zones of thick impasto, while loss of the discolored uppermost layers reveals the light yellow below. The most showy feature of this degradation was the presence of whitish, semitransparent globules that were observed on top of the yellow paint. Cadmium carbonate ($CdCO_3$), cadmium oxide hydroxide,

Journal of Physics: Conference Series **566** (2014) 012021 doi:10.1088/1742-6596/566/1/012021

cadmium carbonate oxide, and cadmium sulfate ($CdSO_4$) are potential CdS photodegradation products identified in the paintings [2]. In view of the fact that the pigment, cadmium sulfide, was historically synthesized by means of dry and wet processes and that $CdCO_3$ and $CdSO_4$ are reagents for this procedure, their identification alone does not constitute conclusive proof of photo-oxidation. This induced the researchers to propose a mechanism where, recurring cycles of moistening and drying could have eventually led to local accumulation of white globules $CdSO_4$*$2H_2O$ at the paint surface after evaporation of the water. This type of degradation was found on the surface of faded cadmium yellow paints in the works of James Ensor (1860-1949) [3] and Henri Matisse[4, 5].
The goal of the current research is to understand the oxidation and hydration mechanisms of these whitish globules in the paint surface by means of Density Functional Theory within the generalized gradient approximation (DFT-GGA). To this end the $\{10.0\}$ surface of CdS was considered as a preliminary benchmark. It was simulated according to the slab geometry and was next covered by oxygen, water molecules and a combination of $O_2$+ $H_2O$. More specifically, we determined the favorite adsorption sites and calculated the adsorption energies of the different molecules on top of the surface. The details of the electronic structure of the interactions are given via the bonding charge analysis along with a thorough description of the geometry.

## 2. Method

All the calculations are performed within the framework of the Density Functional Theory in the Generalized Gradient Approximation (DFT-GGA) as implemented in the quantum-ESPRESSO [6] suite of codes. Ultrasoft pseudopotentials (PBE) are used and van der Waals dispersion forces are accounted for. The Kohn-Sham equations were solved by using a plane waves basis set and the kinetic energy cutoffs for the electron wavefunctions and for the augmented electron density are 28 Ry and 280 Ry, respectively. The Brillouin zone sampling is performed with $4 \times 4 \times 4$ and $4 \times 4 \times 1$ k-points meshes for the bulk and surface structures, respectively. Valence electron configurations included Cd $5s^2$ $5p^2$ $4d^{10}$, S $3s^2$ $3p^4$, O $2s^2$ $2p^4$, and H $1s^1$ states.
The surface energy for the cleavage surface is obtained from:

$$\gamma = \frac{E_{sub} - E_{bulk}}{2A}$$

Where $E_{sub}$ is the total energy of the surface, $E_{bulk}$ represents the total energy of a bulk structure - with the same symmetry - containing as many CdS-pairs as those in the surface, and A is the surface Area. The adsorption energy ($E_{ad}$), always regarded as a measure of the strength of adsorbate-substrate adsorption, is defined as follows:

$$E_{ad} = E_{sub+mol} - [E_{sub} + nE_{mol}]$$

In which $E_{sub+mol}$ is the total energy of the combined system containing the surface and $n$ molecules, $E_{sub}$ is the total energy of the clean surface, and $E_{mol}$ is the energy of a single isolated molecule. The bonding charge analysis, that gives an estimate of the electron charge transfer, was evaluated according to the scheme proposed in Ref.[7], using the following expressions:

$$\Delta\rho(\mathbf{r}) = \rho_{sub+mol} - \rho_{sub} - \rho_{mol}$$

Where $\rho_{sub+mol}(\mathbf{r})$, $\rho_{sub}(\mathbf{r})$ and $\rho_{mol}(\mathbf{r})$ are the charge densities of the whole system, the isolated substrate and the adsorbate, respectively.

## 3. Results and Discussion

The CdS {10.0} surface was cleaved on the basis of the optimised hexagonal (wurtzite) structure along the direction perpendicular to the $[10\bar{1}0]$. The computed lattice parameters for the bulk were a=4.13Å, c= 6.72Å and c/a=1.62 which resulted in very good agreement, within the 0.5%, with the experimental values: $a_{exp}$=4.14Å $c_{exp}$=6.72Å $(c/a)_{exp}$= 1.62 [8]. A slab containing 5 atomic layers was sufficient to represent the CdS {10.0} surface since increasing the number of layers up to 7 layers both the surface energy and the interatomic distances showed no remarkable changes. Due to periodic boundary conditions, the slabs were separated by 20Å of vacuum to avoid interaction between images.

All the atoms were allowed to relax. The structure of the relaxed CdS {10.0} surface showed a characteristic ridged profile: the under-coordinated Cd cations relaxed inward, toward the bulk, whereas the under-coordinated S anions relaxed outward. (see Figure 1). The positions of the atoms in outermost layer are defined in terms of lateral distances *(D)* in either $x$ or $y$ directions parallel to the surface and interlayer distances perpendicular to the surface (⊥).

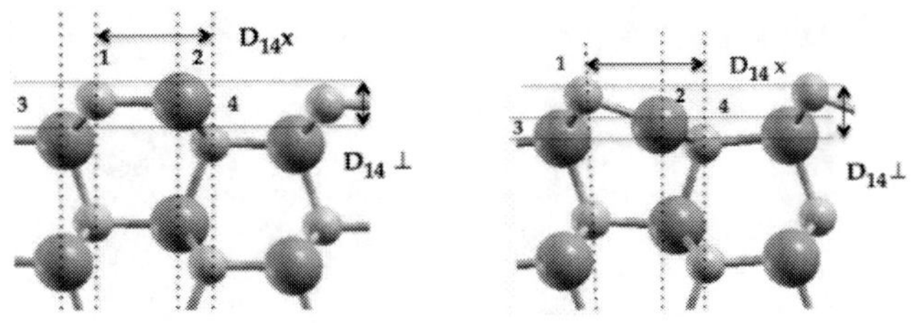


**Figure 1.** Side view of the ideal (left) and relaxed (right) configuration of the CdS {10.0} surface. Distances $D_{1,4}$x and $D_{1,4}$⊥ are meant to be the distances between the atom 1 and the atom 4 along the $x$ direction and the $z$ direction, respectively. The complete set of calculated values is given in Table 1 in comparison with the results of Barnard and collaborators.[9]

In Table 1 are summarised the results of the interatomic distances obtained for the ideal and relaxed clean surface in comparison with the results of Barnard *et al.* [9] obtained with a similar DFT-GGA approach. Also in this case the agreement is excellent, within 1.0% for all the distances but the $D_{1,2}$⊥ that turns out to be 0.02Å shorter than that obtained in Ref. [9]. The discrapancy might be attributed to the difference in the size of the simulation cell. In Ref. [9] the periodic surface has a $2 \times 2$ symmetry versus the $2 \times 1$ of our work.

The calculated surface energy turns out to be $\gamma$ = 0.28 $Jm^{-2}$, which is identical to that obtained in Ref. [9] To simulate the combined effects of exposure to air and humidity, the CdS {10.0} surface was allowed to interact with isolated oxygen, water and then with a combination of $O_2$+$H_2O$. For this specific case, the molecules and the three outermost atomic layers of the surface were allowed to relax, while the two bottom layers were kept fixed to the bulk positions to simulate the infinite crystal. As for the isolated $O_2$ molecule, four setups were considered as initial conditions, where the molecules were differently placed on top of the surface sites prior the relaxation. More specifically, the molecules were initially positioned at a distance as large as 2 Å above the Cd and S sites in both the two non-equivalent positions, namely the *on-top* and *hollow* ones, respectively. In Figure 2, the lowest energy relaxed configuration with $O_2$ located on top the Cd ion is shown. This position corresponds to an adsorption energy as large as -12.99 $kJmol^{-2}$ indicating the molecule to be physisorbed. In this case both the intra- and interlayer distances did not change remarkably while the O-O bond within the oxygen molecule increases from 1.20Å to 1.23Å. The bonding charge analysis reveals an accumulation of charge in the region between $O_2$ and the Cd surface ion (red area) indicating a rearrangement of electron charge around the $O_2$ molecule. The oxygen molecule was then replaced by a water molecule.

**Table 1.** interatomic distances

| CdS | ideal | relaxed | Barnard *et al.*[9] |
|---|---|---|---|
| $D_{1,2}x$ | 2.53 | 2.29 | 2.29 |
| $D_{1,3}x$ | 0.84 | 0.64 | 0.65 |
| $D_{1,4}x$ | 3.37 | 3.50 | 3.50 |
| $D_{1,2}y$ | 0.00 | 0.00 | 0.00 |
| $D_{1,3}y$ | 2.09 | 2.07 | 2.07 |
| $D_{1,4}y$ | 2.09 | 2.06 | 2.07 |
| $D_{1,2}\perp$ | 0.00 | 0.80 | 0.78 |
| $D_{1,3}\perp$ | 1.20 | 1.30 | 1.31 |
| $D_{1,4}\perp$ | 1.21 | 1.55 | 1.57 |

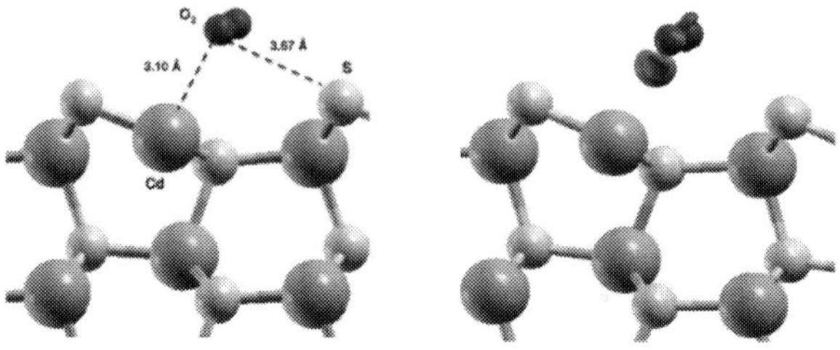


**Figure 2.** Interaction between the CdS {10.0} surface and an isolated $O_2$ molecule: relevant interatomic distances (left) and bonding charge analysis (right)

The result is shown in Figure 3. The $H_2O$ moved toward the hollow-Cd. The adsorption energy, $E_{ad} = -32.09$ kJmol$^{-1}$, is higher than that obtained for $O_2$ and yet, the mechanism suggested is again a physisorption. Also in this system, no evident displacements of the atoms in the substrate are detected. The change of the angle within the water molecule is, conversely, noticeable. It appears to decrease from 104.39° to 103.25°, with are close, within 1%, to the experimental value 104.5° .[10]

Concerning the bonding charge analysis, the water dipole was evident and a slight accumulation (red area) of the charge between oxygen and the surface hollow-Cd resulted in opposition to a depletion of charge (blu area) in the hydrogen regions. Finally, a combination of $O_2$ and

**Figure 3.** Interaction between the CdS {10.0} surface and an isolated $H_2O$ molecule: relevant interatomic distances (left) and bonding charge analysis (right)

$H_2O$ was considered. The two molecules were placed on the surface and let free to relax. The final configuration is reported in Figure 4. The $O_2$ molecule holds the favorite adsorption site above the on-top Cd while $H_2O$ is located on top of the hollow-S and is oriented parallel to the surface with the oxygen atom closer to $O_2$. Again, no atomic position changed within the

**Figure 4.** Interaction between the CdS {10.0} surface and $O_2$ + $H_2O$ molecules: relevant interatomic distances (left) and bonding charge analysis (right)

substrate. The angle $H\hat{O}H$ decreases from 104.39° to 103.25°. A re-arrangement of the charge results in an accumulation in the region between $H_2O$ and S and partially between $O_2$ and Cd. The adsorption energy of this system, $E_{ads} = -46.40$ kJmol$^{-1}$, indicates a physisorption.
The present work has to be regarded as a preliminary study of the combination of elemental molecules interacting with CdS. Further calculations considering different initial conditions and concentrations are in progress.

## 4. Conclusion

The structure and electronic properties of CdS {10.0} surface were studied using a fully theoretical model within the DFT-GGA framework. The adsorption of oxygen and water isolated molecules along with a combination of them was considered. The physical adsorption of the aforementioned molecules on the cleavage surface turns out to be the dominant mechanism. Further investigation considering a defective surface and/or surfaces of different symmetry are to be taken into account.

## Acknowledgments

The authors acknowledge RAS (Regione autonoma della Sardegna) and Fondazione Banco di Sardegna for partial financial support. All the calculations are performed on the HPC infrastructure at CRS4, Pula (Ca), Italy.

## References

[1] Paulus J and Knuutinen U 2004 *Appl. Phys. A* **79** 397-400.
[2] Leone et al 2005 *The deterioration of cadmium sulphide yellow artists' pigments* (London, James & James) p 803-813
[3] Snickt et al 2009 *Anal. Chem.*, **81** 2600-2610
[4] Mass J L, Opila R, Buckley B, Cotte M, Church J, Mehta A 2013 *Appl. Phys. A* Mater. Csi. Process. **111** 59-68
[5] Mass J, Sedlmair J, Schmidt Patterson C, Carson D, Buckley B, Hirschmugl C 2013 *Analyst* **138** 6032-6043
[6] Giannozzi P et al Journal of Physics: Condensed Matter 2009, 21, 395502 (19pp)
[7] Farnesi Camellone M and Fabris S 2009 *JACS articles*
[8] Razik N. 1987 *J. Matter. Sci. Lett.* **6** 1443
[9] Barnard A S and Xu H 2007 *J. Phys. Chem. C* **111** 18112-18117
[10] Eisenberg D and Kauzmann, W 2006 *The Structure and Properties of Water* (Oxford;UK,Claderon Press)

# Advanced imaging systems for diagnostic investigations applied to Cultural Heritage

**E Peccenini[1,2,3], F Albertin[4], M Bettuzzi[1,2,3], R Brancaccio[1,2,3], F Casali[1,2], M P Morigi[1,2,3], F Petrucci[5,6,7]**
[1]Centro Fermi, Roma
[2]Dipartimento di Fisica e Astronomia, Università di Bologna, Italy
[3]INFN sezione di Bologna, Italy
[4]Ecole Polytechnique Fédérale de Lausanne (EPFL), Lausanne, Switzerland
[5]Dipartimento di Fisica e Scienze della Terra, Università di Ferrara, Italy
[6]INFN sezione di Ferrara, Italy
[7]Laboratorio TekneHub, Tecnopolo di Ferrara, Italy

E-mail: eva.peccenini@unibo.it

**Abstract.** The diagnostic investigations are an important resource in the studies on Cultural Heritage to enhance the knowledge on execution techniques, materials and conservation status of a work of art. In this field, due to the great historical and artistic value of the objects, preservation is the main concern; for this reason, new technological equipment has been designed and developed in the Physics Departments of the Universities of Ferrara and Bologna to enhance the non-invasive approach to the study of pictorial artworks and other objects of cultural interest. Infrared (IR) reflectography, X-ray radiography and computed tomography (CT), applied to works of art, are joined by the same goal: to get hidden information on execution techniques and inner structure pursuing the non-invasiveness of the methods, although using different setup and physical principles. In this work transportable imaging systems to investigate large objects in museums and galleries are presented. In particular, 2D scanning devices for IR reflectography and X-ray radiography, CT systems and some applications to the Cultural Heritage are described.

## 1. Introduction

Works of art and archaeological objects constitute the main part of the tangible Cultural Heritage, they have the common features of uniqueness and irreplaceability; for these reasons, they are often kept in museums or galleries for preservation purposes and they can hardly be moved.

These objects of cultural interest are also extremely different from each other, characterized by different materials and sizes, therefore, industrial and medical diagnostics apparatus cannot be used for applications in this field. Developing custom imaging devices becomes essential to meet specific needs as: non-invasiveness, transportability and versatility.

The devices presented in this paper have been designed and realized by two research groups: the X-ray Imaging group at the Department of Physics and Astronomy of Bologna University and the Archaeometry group at the Department of Physics and Earth Science of Ferrara University. These two groups have been involved for several years in the development of non-invasive experimental apparatus using X-rays and other types of radiation for applications on cultural heritage.

## 2. Scanning device for wide band infrared reflectography

IR reflectography is a non-invasive image diagnostic technique widely used to reveal underdrawing in paintings. It benefits from the property of near-IR radiation (700-2500 µm in wavelength) to be less scattered by pictorial layers than shorter wavelengths. Consequently it may often reach the deepest layers, hidden by superimposed colours and bring back the information stored there. The reflectographic capability depends on the optical properties of the pictorial layer and on the characteristics of the detector [1] [2].

The main devices used to this aim have been: Vidicon Tube, Si CCD cameras and scanning systems. Vidicon tube, with a spectral response up to 2.2 µm was for over 20 years, the only choice for IR Reflectography. The spectral response of Vidicon tube allowed a good detection of underdrawing, but it had low imaging performance and high cost. During the 90's, the Vidicon tube was replaced by the Silicon CCD infrared camera more handy and cheaper, but with spectral response limited at 1.1 µm. The need of high-quality optical images and wider spectral response led, in the 90's, to the development of IR scanning systems [3]. The device presented here has a spectral response extended up to 2.5 µm to improve reflectography capability to read the underdrawing in paintings [4].

### *2.1. Scanning device and optical apparatus*

The scanning device, developed at the Department of Physics and Earth Science in Ferrara University, is composed of two motorized linear guides, supported by a structure of aluminum profiles. The image sampling is acquired during the horizontal movement, 90 cm wide, performed at 0.2 m/sec mean speed. Signals from the photodiode are acquired with a horizontal pitch of 5 µm, at 12 bit/pixel.

The optical system is a binocular microscope head mounted on a Y-axis, provided by a revolver with selectable magnification steps. The choice of the microscope is justified to enlarge the area focused on the detector to cover the active area of photodiode, to warrant a working distance of about 10 cm, suitable to avoid damage to the painting, and to allow a depth of field of about 0.5 cm, compatible with the roughness of pictorial surfaces.

The lighting system includes two halogen lamps, placed at 45°, equipped with a quartz lens ($f = 850$ mm), to concentrate the light to the target area, and a silicon filter to cut wavelengths shorter than 1100 nm, to avoid heating of the painting by non-IR radiation.

A Two-color© detector by Hamamatsu with thermoelectric cooling is used. It contain a Si photodiode (spectral range from 0.27 to 1.1 µm), mounted over an InGaAs PIN photodiode (spectral range from 0.94 to 2.57 µm), along the same optical axis. Due to transparency of Silicon at wavelength over 1.1 µm, this detector delivers the possibility of acquiring two images simultaneously in different spectral bands.

However, for applications of wide band reflectography the InGaAs detector is used. With 4x magnification the InGaAs photodiode has a resolution of 2 lp/mm [4].

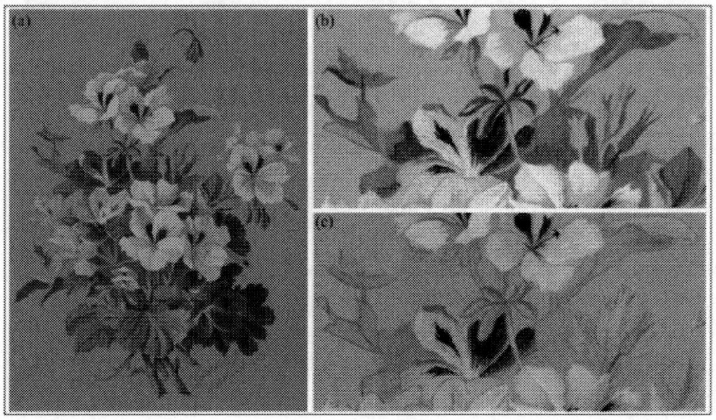

**Figure 1.** (a) G. Nardozzi, Flowers, 1889, oil on paper, private collection. (b) Detail of reflectographic image acquired by Si detector.
(c) Detail of reflectographic image acquired by InGaAs detector [4].

### *2.2. Examinations of paintings*

The application shows the comparison between Si and InGaAs detector to highlight the better readability of the underdrawings using a spectral band extended to longer wavelength. The painting, showed in Figure 1 (a), has thin and uniform layer of not treated paper. The artist reproduces the flowers in a very realistic way, like in a herbarium, typical features of academic approach characterized by a fine technique.The infrared reflectography with Si detector discloses the underdrawing along the edge of the white petals, where the paint layer is thinner, and under the light green of the leaves, (Figure 1 b). Instead, the dark green appears more absorbent. In the image acquired by InGaAs detector, the underdrawing is visible under both green pigments. Moreover, the drawing under the white pigment, hardly recognizable by Silicon detector, appears here well clear (Figure 1 c) [4].

## 3. X-ray digital radiography: comparison of two devices

X-ray radiography provides a wealth of information not obtainable with other imaging techniques, for its capability to cross all layers of a painting.
X-ray imaging has evolved over time. In digital radiography, semiconductor detectors have replaced the traditional photographic film. In Cultural Heritage field this transition has been slowest for two basic reasons: the need of handy and lightweight devices for application on site and the high cost of digital detectors. Therefore, an intermediate solution of photostimulable phosphor plates are still widely used. The radiographic data are captured and stored in the phosphor plates, and then, they are converted in a digital image using a scanner by laser stimulation. This system warrant a good dynamic range, but, unlike digital detectors, does not provide an immediate availability of the radiographic image.

The digital detectors such as CCD, CMOS or flat panel allow a real-time image capture but need motorized devices to warrant the movement during the acquisition.

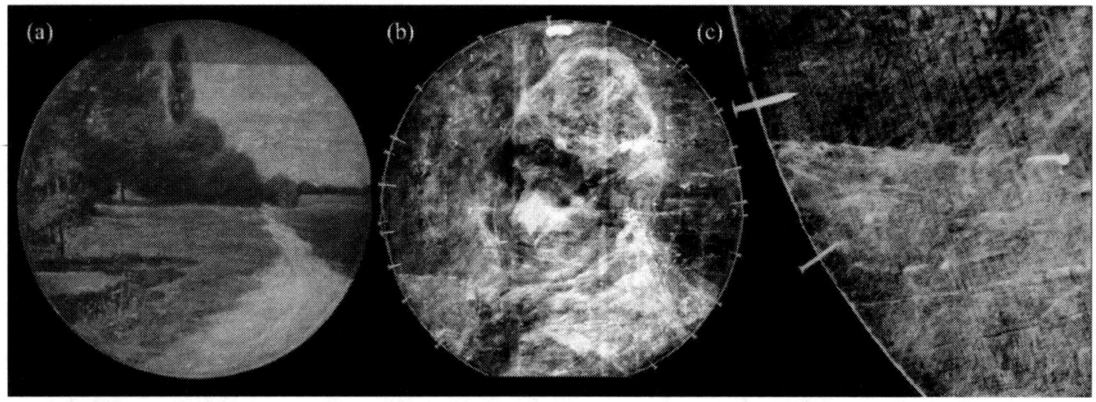


**Figure 2.** (a) Landscape , oil on wood, 20th century, private collection. (b) Digital X-ray radiography, (c) detail of radiography [5].

The radiography of the landscape in Figure 2 has been acquired at 30 kV using a CMOS detector mounted on a scanning device. The radiographic image reveals the presence of a woman portrait hidden under the pictorial layers. Due to dynamic range of 12-bit digitalization, the visualization of all details of the painting is performed by adjusting contrast and brightness settings.

In Figure 2 c elements with high absorption as nails inside the support are visible, but in the same image, details of low absorption, like as the outer edge, probably made of leather or cardboard, and the canvas applied on the support are also visible [5].

Two scanning devices for digital radiography with different setup are presented: a transportable X-ray scanning device for on-site applications and an X-ray scanning device for large paintings

The main difference between the two systems is the distance between the source and the detector: In the transportable system the source is independent by the detector movement, conversely, in the system for large paintings the X-ray tube is at a fixed distance of about 60 cm and moves with the detector during the scan, this allows to use a low current and shorter exposure time, but a smaller focal spot of the X-ray source is required (table 1).

The sources used have different kVp, but typically 45 kV are enough for the thicker paintings on wood, such as the altarpieces, however, in these cases the transportable system are preferable to perform in situ analysis. On the contrary, the system for large paintings is suitable for applications on paintings on large canvas for the ability to perform acquisition on larger areas (up to 2 $m^2$) and for the possibility to use a lower voltage (up to 20 kV).

The choice of the detectors was based on the pixel size, in order to be able to distinguish *craquelure* of at least 100 μm. Both detectors are indirect conversion and contain a two-dimensional CMOS photodiode array. They have similar resolution, but the Hamamtsu detector mounted on the transportable device has a larger field of view, 12 x 12 $cm^2$ instead of 10 x 10 $cm^2$ of Rad eye detector (table 1).

In both devices the tile scanning technique is performed. The radiographic images are acquired according to a grid that warrants an overlap of few centimetres between the images. In the processing phase, several commercial or open source software are used to merge the single images, such as PTGui, PanaVue ImageAssembler, ImageJ or Adobe Photoshop.

**Table 1.** Main characteristics of sources and detectors used in the devices

| **Devices** | X-ray tubes | Detectors |
|---|---|---|
| Transportable X-ray scanner | Gilardoni MHF 200D, tungsten anode, high voltage range 30÷200 kV, current range 1÷8 mA, max power 900 W, beryllium window, focal spot 1.5 x 1.5 mm. | Hamamatsu C10900D, CsI scintillator, Pixel size 100x100 μm, active area 12x12 $cm^2$ (1216x1232 pixels), 16-bit digital output. |
| X-ray scanner for large paintings | Varian M147, molybdenum anode, high voltage range 20÷50 kV, max current 15 mA, beryllium window, nominal focal spot 0.1 – 0.3 mm. | Rad eye 200, GOS scintillator, Pixel size 96x96 μm, active area 10x10 $cm^2$ (1024x1000) pixels, 16-bit digital output. |

### *3.1. Transportable X-ray scanning device for on-site applications*

In the Department of Physics and Astronomy at University of Bologna a scanning device easy to disassemble and transport has been developed specifically for *in situ* applications. The system is composed of a pair of orthogonal lightweight and decomposable axes, each consisting of a single aluminium profile. The axes allow the translation of the detector, with a stroke of 1.5 m. The X-ray tube is separated from the structure and is placed opposite to painting at a proper distance in order to irradiate the whole scanned area (figure 3). In this system, the high-energy source has been chosen for a further upgrade for tomographic applications. The maximum size of detectable area is 1.5 x 1.5 $m^2$ with resolution of 100 μm.

### *3.2. X-ray scanning device for large paintings.*

The scanner allows the investigation of paintings of size up to 2 x 2 $m^2$ with a spatial resolution of 96 μm, it has been designed and realized at the Larix Lab of Department of Physics and Earth Science in Ferrara University. The system can be disassembled and moved to perform *in situ* X-ray scan.

The X-ray tube and the detector are mounted specularly on a double aluminium motorized arm. This setup provides the best detector-source stability. The painting is locked in a central frame, while the X-ray tube and the detector are moved (Figure 4). Movement, emission and acquisition are synchronized [6].

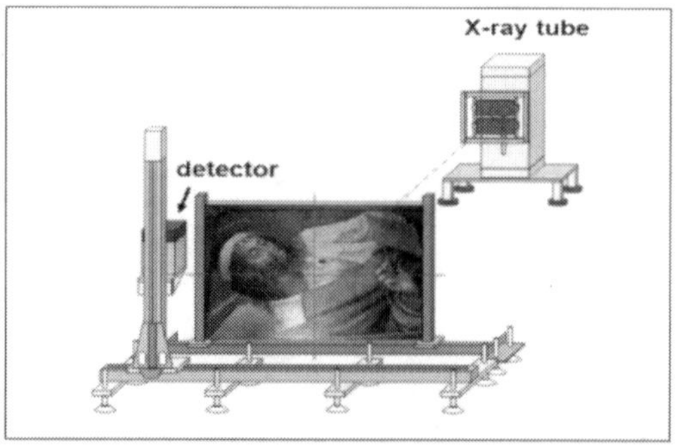


**Figure 3.** Transportable X-ray scanning device.

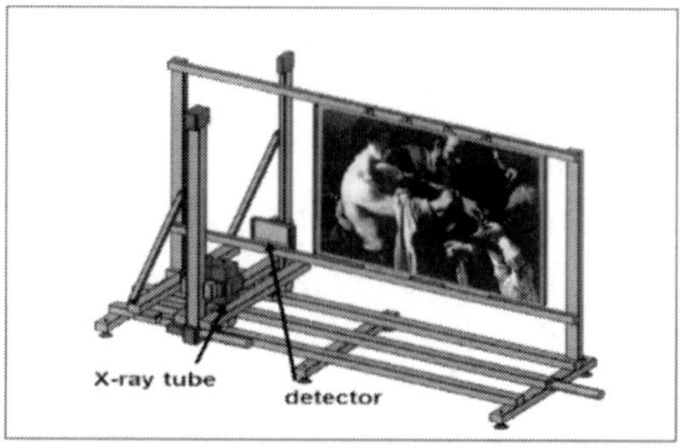


**Figure 4.** X-ray scanning device for large paintings.

## 4. Computed tomography systems

Tomographic systems, routinely used in medical or industrial field, consist in stationary devices intended to investigate objects, or patients, very similar to each other in size, structure and composition.

This technique requires, for applications on cultural heritage, the development of tomographic systems versatile and transportable, able to carry out *in situ* investigations, where the works of art or archaeological finds are preserved.

To make a CT system, three items are required: an X-ray source, a detector and a rotating table.

The data acquisition phase requires the rotation of the object around a fixed axis. The object is mounted on a support locked on the rotating table, which moves with high precision by means of a control software. At every rotation step, a radiographic image is acquired. In this process an X-ray projection of the object is projected on the detector and the geometry of the system source-object-detector defines its size. The processing of the set of data with a reconstruction procedure and specific algorithms allows the reconstruction of the external and internal structure of the object [7]. Below is a description of two X-ray CT systems developed at the Department of Physics and Astronomy of Bologna University and at the Te.Co.Re Lab.

### *4.1. Transportable X-ray CT system for small-medium size object*

In the X-ray Imaging Lab at the Department of Physics and Astronomy of Bologna University a CT system has been realized for applications on small-medium objects. This system uses an X-ray tube with a maximum voltage of 120 kV and an indirect conversion flat-panel detector (with structured cesium iodide scintillator) of 25 x 20 $cm^2$, mounted on two orthogonal translation axes which extend the area up to 50x50 $cm^2$. This detector allows to perform CT with a good resolution of 127 μm, it is particular useful for applications on bones remains. The lightweight of components give the system easily transportable for application on-site.

### *4.2. X-Ray CT system for large objects*

A CT equipment for application on large objects is set up at Te.Co.Re. Lab located in Ravenna. It is equipped with an X-ray tube with energy up to 200 keV. The detector consists of a box containing a CCD camera (2184×1472 pixels) with interchangeable lens, a mirror, and a structured cesium iodide scintillator screen (45x45 $cm^2$ size and 1mm thick). Part of the X-rays emitted by the source passes through the object and reaches the scintillator screen, which converts radiation to visible light. The mirror, placed in the box at 45°, reflects the light from the screen to the CCD camera for acquisition of the radiographic projections.

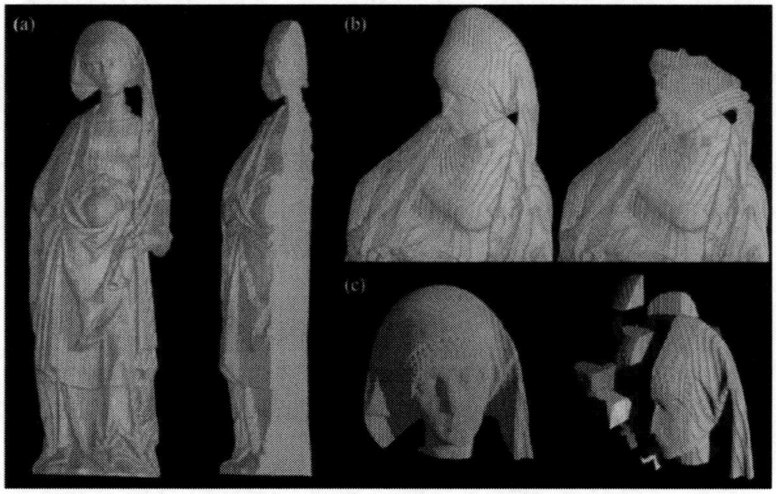


**Figure 5.** F. Deflorian, 2012, wooden statue, 90 x 30 cm, private collection. 3D reconstruction of tomographic data. (a) Whole and axial section, (b) sagittal section of the bust, (c) head and its unconventional rendering.

The detection system has a maximum resolution of 132 um, but it is characterized by presence of outlier artifact due to the scattered radiation that reach the ccd camera, therefore, generally the CT are carried with a resolution of 264, applying a binning 2x2 in the acquisition phase, to reduce the noise and the acquisition time.

The active area of the detector is 45x45 $cm^2$ which can be extended up to almost 3.5 x1.0 $m^2$ using a mechanical axis with horizontal stroke of 3 meters and a vertical lift of 70 cm [7].

The wide field of view allows the use of the system for CT on objects of 1 - 2 m in size. On the contrary of the other, this system is used only in the laboratory.

Figure 5 show an application to a wooden statue of contemporary art. It is the 3D reconstruction obtained from tomography data set. Scrolling the slices of the volume is possible to inspect the inner part of the object. In the axial and sagittal sections (Figure 5 a, b) the different blocks of wood that compose the statue have been identified. Through a rendering tool, the single blocks have been split to highlight the structure of the artwork (Figure 5 c).

## 5. Conclusions

The growing interest of museums, superintendencies and restoration centers in image diagnostic has brought several groups to focus their research on new systems for application on Cultural Heritage. The collaboration among institutions and research centers, such as Universities, INFN and TekneHub has allowed the realization of the systems presented in this paper.

All the systems described are an evolution of traditional systems. The scanning system for IR reflectography has been developed to fulfill the need to increase the spectral range and eliminate the aberrations of the images. The X-ray systems, already widely used in medical and industrial applications, have been adapted to the cultural heritage applications. They have to be brought where the works of art are kept, with setup dedicated to objects of different size and appropriate energy range to investigate different materials.

The next goal of the X-ray Imaging Group of the University of Bologna, with the collaboration of INFN and the Fermi Centre of Rome, will be the construction of a higher energy CT system using a 320 kV X-ray tube to extend applications to objects denser or thicker, than those currently accessible.

Journal of Physics: Conference Series **566** (2014) 012022 doi:10.1088/1742-6596/566/1/012022

**References**

[1] Walmsley E, Metzger C, Delaney J K, Fletcher C 1994 *Stud. Conserv.* **39** 217
[2] Gargano M, Ludwig N, Poldi G 2007 *Infrared Phys. Technol.* **49** 249
[3] Bertani D, Cetica M, Poggi P, Puccioni G, Buzzegoli E, Kunzelman D, Cecchi S 1990 *Stud. Conserv.* **35** 313
[4] Peccenini E 2012 *A scanning device for wide band infrared reflectography* PhD thesis (Ferrara: Università degli studi di Ferrara Eprints)
[5] Albertin F 2011 *K-edge radiography and applications to Cultural Heritage* PhD thesis (Ferrara: Università degli studi di Ferrara Eprints)
[6] Albertin F, Boselli L, Chiozzi S, Peccenini E, Pellicori V, Petrucci F, Tisato F 2013 *Progetto Restauro* (Padova: Il Prato) **65** 20, 40
[7] Morigi M P, Casali F, Bettuzzi M, Brancaccio R, D'Errico V 2010 *Appl. Phys.* **100** 603

5th Young Researcher Meeting, Trieste 2014 IOP Publishing
Journal of Physics: Conference Series **566** (2014) 012023 doi:10.1088/1742-6596/566/1/012023

# Dynamical Bridges: The Electromagnetic Case

**Eduardo Bittencourt**

CAPES Foundation, Ministry of Education of Brazil - Brasília - Brazil and Sapienza Università di Roma - Dipartimento di Fisica, P.le Aldo Moro 5 - 00185 Rome - Italy

E-mail: eduardo.bittencourt@icranet.org

**Abstract.** We present the main line of the Dynamical Bridge method applied to electromagnetic (EM) fields. In this framework, we show the equivalence between the Born-Infeld theory written in a given curved space to the Maxwell's one written in the flat space. In the limit of weak EM fields, we obtain a geometrical contribution to the anomalous magnetic moment of the leptons and, finally, we compare it with experimental data.

## 1. Introduction

Recently, we have studied a mathematical map called "Dynamical Bridge" (DB) between distinct dynamical equations written in different background metrics (see Ref. [1] and references therein). This map corresponds to a new way of describing dynamical equations for fundamental fields, because it gives more than one representation of the same thing. In other words, the approach changes the space-time metric and the differential operator acting upon the dynamical field such that it is kept unchanged. Therefore, we are able to hide nonlinear terms into the curvature of the space-time.

Following Poincaré's viewpoint [2] concerning the relationship between Geometry and Physics, we consider a curved metric without any gravitational character. Namely, the curved geometry introduced here is seen only as a way to describe geometrically nonlinear interacting terms. In particular, we specify the geometries such that the metric and its inverse assume the binomial form

$$\widehat{q}_{\mu\nu}(x^\alpha) = A(x^\alpha)\,\eta_{\mu\nu} + B(x^\alpha)\,\Phi_{\mu\nu}, \qquad \widehat{q}^{\mu\nu}(x^\alpha) = \alpha(x^\alpha)\,\eta^{\mu\nu} + \beta(x^\alpha)\,\Phi^{\mu\nu}, \tag{1}$$

where $\mu, \nu$ are space-time indexes running from 0 to 3 and $x^\alpha$ represents space-time points. The functions $A$, $B$, $\alpha$ and $\beta$, the Minkowski metric $\eta^{\mu\nu}$ and the symmetric tensor $\Phi^{\mu\nu}$ are subjected to

$$\Phi_{\mu\nu}\,\Phi^{\nu\lambda} = m\,\delta^\lambda_\mu + n\,\Phi^\lambda_\mu. \tag{2}$$

where $\delta^\lambda_\mu$ is the Kronecker delta. Note that $A$ and $B$ are completely determined if $m$, $n$, $\alpha$ and $\beta$ are known. This kind of metric can be seen as a disformal transformation [3] of the reference metric, which was chosen to be the Minkowski one.

The applications of this type of transformation lead us to alternative explanations of several phenomena for different fields which are not yet understood from the standard model viewpoint. For instance, we construct from scalar fields a geometric scalar theory of gravity [4] with $\Phi_{\mu\nu}$ given in terms of gradients of the fundamental field; for massless spinor fields we could explain

Journal of Physics: Conference Series **566** (2014) 012023 doi:10.1088/1742-6596/566/1/012023

the chiral symmetry breaking [5] assuming $\Phi_{\mu\nu}$ as a tensorial product of linear combinations of the spinorial currents; for EM fields we derive a geometrical contribution to the magnetic moment of leptons [1], where $\Phi_{\mu\nu}$ is given in terms of Faraday tensor.

We will focus here only on the EM case. We start with an application in the kinematical context, explaining how the DB can be used to describe accelerated motions in a given space as geodesics in another space. Then, we move to the dynamical context and we construct the DB for EM fields in a very simple way, where the physical aspects can be discussed with more details. In the last part of the paper we introduce a comparison with experimental data.

## 2. Kinematical case

We present an extension of previous works [6, 7] concerning the DB in the kinematical context. We find a general and compact formula for the main theorem, which is valid now for any background (not only the Minkowski one) and it is independent of the functional form of $\widehat{q}_{\mu\nu}$:

**Theorem:** *Consider a congruence of curves represented by the vector field $V_\mu$ in a given space-time and parameterized by the affine parameter $\lambda$.*[1] *This vector follows along a geodesic in the metric $\hat{q}_{\mu\nu}$ if the condition is satisfied*

$$\frac{1}{2}\hat{N}_{,\mu} + V_{[\mu,\nu]}\hat{q}^{\nu\alpha}V_\alpha - p(\lambda)V_\mu = 0, \tag{3}$$

*where comma means partial derivative w.r.t the coordinates $x^\mu$ and $V_{[\mu,\nu]} \equiv V_{\mu,\nu} - V_{\nu,\mu}$. The norm $\hat{N}$ of $V_\mu$ is calculated with the metric $\hat{q}_{\mu\nu}$ and $p(\lambda)$ represents the parametrization chosen.*

Applying this for a special set of metrics obtained from the disformal transformations [3], which depend on the background and the normalized field $V_\mu$ under consideration, that is

$$\hat{q}^{\mu\nu} = \alpha\eta^{\mu\nu} + \beta V^\mu V^\nu, \quad \text{and inverse} \quad \hat{q}_{\mu\nu} = \frac{1}{\alpha}\eta_{\mu\nu} - \frac{\beta}{\alpha(\alpha+\beta)}V_\mu V_\nu, \tag{4}$$

one can describe in terms of geodesics the motion produced by any conservative force. For instance, suppose that $V_\mu$ has an acceleration $a_\mu \equiv \partial_\mu\Phi$ in the Minkowski space, where $\Phi$ is a scalar potential. The condition for $V_\mu$ to follow along a geodesic in the metric (4) is

$$a_\mu = -\frac{1}{2}\frac{(\alpha+\beta)_{,\mu}}{\alpha+\beta}. \tag{5}$$

It implies that $\Phi$, $\alpha$ and $\beta$ are related through

$$\alpha + \beta = e^{-2\Phi}.$$

## 3. Dynamical Context: Application to Vector Fields

The electromagnetic DB we present here corresponds to the equivalence between the Born-Infeld (BI) theory written in a curved space and the Maxwell theory expressed in flat space.

Let us start considering the BI Lagrangian [8]

$$L = \beta^2\left(1 - \sqrt{\hat{U}}\right), \tag{6}$$

where $\hat{U} \equiv 1 + \hat{F}/(2\beta^2) - \hat{G}^2/(16\beta^4)$ and $\hat{F} \equiv F_{\mu\nu}F_{\alpha\beta}\hat{e}^{\mu\alpha}\hat{e}^{\nu\beta}$ and $\hat{G} \equiv F_{\mu\nu}F^*_{\alpha\beta}\hat{e}^{\mu\alpha}\hat{e}^{\nu\beta}$ are the two independent scalars constructed with the 'electromagnetic' metric $\hat{e}^{\mu\nu}$ and the Faraday

[1] In particular, $\lambda$ can be seen as the arc length of the curve.

tensor $F_{\mu\nu}$. The free parameter $\beta$ corresponds to the critical field of the BI theory. The curved space-time is assumed to be

$$\hat{e}^{\mu\nu} \equiv a\,\eta^{\mu\nu} + b\,\Phi^{\mu\nu}, \quad \text{with} \quad \Phi_{\mu\nu} \equiv F_{\mu\alpha}\,F^{\alpha}{}_{\nu}. \tag{7}$$

The variational principle applied to (6) yields the following dynamical equations

$$\frac{1}{\sqrt{-\hat{e}}}\partial_\nu \left[\frac{\sqrt{-\hat{e}}}{\hat{U}}\left(\hat{F}^{\mu\nu} - \frac{1}{4\,\beta^2}\,\hat{G}\,\hat{F}^{*\mu\nu}\right)\right] = 0, \tag{8}$$

where $\hat{e} = \det(\hat{e}_{\mu\nu})$. We also consider the Maxwell Lagrangian defined in the Minkowski space

$$L = -\frac{1}{4}F, \tag{9}$$

with $F \equiv F_{\mu\nu}\,F_{\alpha\beta}\eta^{\mu\alpha}\eta^{\nu\beta}$. The dynamical equations in this case are

$$\frac{1}{\sqrt{-\eta}}\partial_\nu\left(\sqrt{-\eta}F^{\mu\nu}\right) = 0. \tag{10}$$

where $\eta = \det(\eta_{\mu\nu})$. To demonstrate their equivalence is necessary to know how $\hat{F}^{\mu\nu}$ and its dual are related to the Maxwell's fields. Their correspondence is expressed in terms of two algebraic equations

$$n - \epsilon F m + \frac{\hat{G}}{2\beta^2}\epsilon m = -\frac{Q}{4}, \quad \text{and} \quad -\epsilon G m + \frac{\hat{G}n}{2\beta^2} = 0, \tag{11}$$

where $\epsilon \equiv b/a$ and $n = 1 + \frac{1}{16}\epsilon^2 G^2$, $m = 1 - \frac{1}{4}\epsilon F$ and $Q = 1 - \frac{1}{2}\epsilon F - \frac{1}{16}\epsilon^2 G^2$.

The general solution of these equations exists, but it is very complicate. Thus, we shall consider the case of weak EM fields, i.e., only first order corrections in $F \ll \beta^2$. To simplify the second Eq. of (11), we assume the particular configuration for the EM field in which $G = 0$.[2] Then, without loss of generality, we choose $a = 1$.

In this regime, the coefficient $\epsilon$ has a unique physical solution

$$\epsilon = \frac{2}{F}\left(1 - \frac{1}{\sqrt{1 - F/2\beta^2}}\right)$$

and, finally, the EM metric becomes

$$\hat{e}_{\mu\nu} \approx \eta_{\mu\nu} + \frac{1}{2\,\beta^2}\,\Phi_{\mu\nu}. \tag{12}$$

Now, we are ready to introduce an interaction between the EM field and leptons in the curved space, looking for nontrivial results in the flat space. In this context, the interaction is guided by some principles

(i) *Universality*: all particles (charged or not) interact with $\hat{e}_{\mu\nu}$.

(ii) The interaction is done through the *minimal coupling principle*.

Of course the Clifford algebra, the intrinsic algebra of spinors, must be preserved in both space-times. Then, from assumption (i) and Eq. (12) we obtain [1]

$$\hat{\gamma}^\mu = \gamma^\mu - \frac{1}{4\,\beta^2}\Phi^\mu{}_\alpha\gamma^\alpha. \tag{13}$$

where $\gamma_\mu$ are the Dirac matrices. It should be remarked that the first hypothesis (*universality*), commonly used in geometric theories of gravity, implies here a coupling between uncharged particles and the EM field. This will be crucial for coupling with neutrinos as we shall see later.

[2] This condition implies that field configurations like plane waves are solutions of both dynamics.

## 4. Geometric magnetic moment for leptons

In the curved space the dynamics of a spinor $\Psi$ is

$$i\hbar c\hat{\gamma}^{\mu}\hat{\nabla}_{\mu}\Psi - mc^2\Psi = 0, \tag{14}$$

where $\hat{\nabla}_{\mu} \equiv \partial_{\mu} - \hat{\Gamma}^{FI}_{\mu} - \hat{V}_{\mu}$, $\hat{\Gamma}^{FI}_{\mu}$ is the Fock-Ivanenko connection and $\hat{V}_{\mu}$ is an arbitrary element of the algebra given by

$$\hat{V}_{\mu} = i\frac{mc}{\hbar}\frac{F_{\mu\nu}}{\beta}\hat{\gamma}^{\nu}\gamma_5, \tag{15}$$

where $\gamma_5 = (i/4!)\epsilon_{\alpha\beta\mu\nu}\gamma^{\alpha}\gamma^{\beta}\gamma^{\mu}\gamma^{\nu}$ and $\epsilon_{\alpha\beta\mu\nu}$ is the Levi-Civita tensor. This yields $\hat{\nabla}_{\mu}\hat{\gamma}^{\nu} = [\hat{V}_{\mu}, \hat{\gamma}^{\nu}]$, but keeps the metricity condition $\hat{\nabla}_{\alpha}\hat{e}^{\mu\nu} = 0$ and preserves the current $\hat{\nabla}_{\mu}\hat{J}^{\mu} = 0$ (see details in Ref. [1]).

Using the expressions for $\hat{e}_{\mu\nu}$, $\hat{\gamma}^{\mu}$ and $\hat{V}_{\mu}$, we obtain the following equation for $\Psi$ in Minkowski space

$$i\hbar c\gamma^{\mu}\partial_{\mu}\Psi + \frac{mc^2}{2\beta}F_{\mu\nu}[\gamma^{\mu}, \gamma^{\nu}]\gamma_5\Psi - mc^2\Psi = 0. \tag{16}$$

Note that this equation provides a magnetic moment for $\Psi$, whose intensity is

$$\mu^G = \frac{mc^2}{\beta}. \tag{17}$$

We name this contribution as *geometric magnetic moment*.

Charged particles have a classical source for the magnetic moment $\mu$. For instance, electrons have the Bohr magneton $\mu_B = e\hbar/2m_e$. Therefore, the total value of the electron magnetic moment should be read as

$$\mu_e = \mu_B + \frac{m_e c^2}{\beta} + \text{quantum corrections}.$$

The first term is classical, the second one is the geometric magnetic moment and the last contribution represents the quantum corrections due to the presence of charge.

## 5. Comparison with experiments

The anomaly $a_l$ of the magnetic moment $\mu_l$ of a given particle $l$ corresponds to a measurable discrepancy between the theoretical prediction and the experimental result. It is defined as

$$a_l = \frac{g_l - 2}{2} = \frac{m_l}{m_e}\frac{\mu_l}{\mu_B} - 1, \tag{18}$$

where $l = (e^-, \mu, \tau)$, $g_l$ is the Landé factor and $m_l$ is the mass. There are measurements of this anomaly[3] for the electron $e^-$ (the most precise measurement) and for the muon $\mu$ (test of the consistency of the standard model). For $\tau$'s, the anomaly is unobservable due to its short mean-life ($\sim 2.9 \times 10^{-13}$ s) but predictions indicate that it is less than $1.3 \times 10^{-2}$.

$\mu$-case: Due to the particular role played by the anomaly of muons, the difference $\Delta a$ between the $a_l$-experimental and $a_l$-theoretical is assumed to provide the upper limit in which the geometric magnetic moment $\mu^G$ can contribute. So, if the effects of modifications caused by the existence of $\hat{e}^{\mu\nu}$ appear at this order, then we get

$$\mu^G_{\mu} \doteq \Delta a_{\mu}\frac{e\hbar}{2m_{\mu}}. \tag{19}$$

[3] All experimental data used here were taken from Particle Data Group [9]. In particular, in this section we use $\mu_B = 5.788 \times 10^{-11}\text{MeV T}^{-1}$.

On the other hand, the formula of $\mu^G$ is given by

$$\mu_\mu^G = \frac{m_\mu c^2}{\beta}. \tag{20}$$

Assuming that $\mu_\mu^G$ corresponds to the remaining value of muon's anomaly, then we can estimate the critical field as $\beta \approx 1.31 \times 10^{23}\, T$.

$e^-$-case: Assuming this value for $\beta$, the geometric contribution for the electron is

$$\mu_e^G = 6.74 \times 10^{-14} \mu_B \tag{21}$$

and the discrepancy measured between theory and experiments is

$$\Delta a_e = -0.40 \times 10^{-12}. \tag{22}$$

Note that theory and experiments are in agreement up to few $10^{-12}$. For smaller scales the error bars are important, so that the geometric correction $\mu_e^G/\mu_B \sim 10^{-14}$ is not observable for $e^-$.

$\nu$-case: Since neutrinos have zero charge, it is expected that their magnetic moment $\mu_\nu$ and the corresponding quantum corrections are zero. However, the extended standard model suggests that $\mu_\nu \neq 0$ and $\mu_\nu \propto m_\nu$. With this in mind, it seems natural to propose that $\mu_\nu$ has only the geometrical origin which would be

$$\mu_\nu \equiv \frac{m_\nu c^2}{\beta} = 1.32 \times 10^{-19} \mu_B. \tag{23}$$

This value is in agreement with phenomenology [10, 11, 12] and close to the extended standard model prediction ($\sim 10^{-19}\mu_B$).

## 6. Conclusion

The geometrical description of the nonlinearity of the Born-Infeld theory allowed us to explain, through the universality of the metric and the minimal coupling principle, part of the anomalous magnetic moment of leptons and, in particular, of neutrinos. The comparison to experimental data indicates that our proposal is in agreement with observations. So, this model should be deeply investigated in order to improve our understanding about the role of curved geometries in Physics and not only in the gravitational context. This is currently under analysis.

## Acknowledgments

The author thanks Prof. M. Novello and the anonymous referee for the useful comments and the CAPES-ICRANet program (BEX 13956/13-2) for the financial support.

## References

[1] Novello M and Bittencourt E 2014 *Int. J. Mod. Phys.* A **29** 1450075
[2] Poincaré H 1905 *Science and Hypothesis*, (New York: The Walter Scott Publishing CO.)
[3] Bekenstein J D 1993 *Phys. Rev.* D **48** 3641
[4] Novello M, Bittencourt E, Moschella U, Goulart E, Salim J M and Toniato J D 2013 *J. Cosmol. Astropart. Phys.* JCAP06(2013)014
[5] Bittencourt E, Faci S and Novello M 2014 *Int. J. Mod. Phys.* A **29** 1450145
[6] Novello M and Bittencourt E 2012 *Phys. Rev.* D **86** 124024
[7] Novello M and Bittencourt E 2013 *Gen. Rel. Grav.* **45** 1005
[8] Born M and Infeld L 1934 *Proc. R. Soc. A* **144** 425
[9] Beringer J *et al.* 2012 Particle Data Group *Phys. Rev.* D **86** 010001
[10] Kuznetsov A V and Mikheev N V 2007 *J. Cosmol. Astropart. Phys.* JCAP11(2007)031
[11] A.B. Balantekin 2006 *Proc. Origin of Matter and the Evolution of Galaxies (OMEG05)* vol 847 ed Kubono S, Aoki W, Kajino T, Motobayashi T and Nomoto K (Tokyo: AIP Conference Proceedings)
[12] Lychkovskiy O and Blinnikov S 2010 *Phys. Atom. Nucl.* **73** 614

# Gravity and thermodynamics: fundamental principles and gravothermal instability

**G Fragione**[1,2]

[1]Department of Physics, University of Rome La Sapienza, Piazzale Aldo Moro 2, 00185, Rome Italy

[2]Department of Physics, University of Rome Tor Vergata, Via Orazio Raimondo 18, 00173, Rome Italy

E-mail: giacomo.fragione90@gmail.com

**Abstract.** Globular clusters are stellar systems in which thermodynamics plays a fundamental role. Classically these systems are described by non-relativistic single mass King distribution function, which derives from the Fokker-Planck equation by taking into account stellar collisions and host galaxy tidal confinement. In terms of statistical mechanics the available phase space is restricted and so thermodynamics variables assume different meaning respect to the Boltzmann distribution. Through the introduction of an effective potential, which stands for the galactic tidal forces, a new thermodynamic theory is developed and theoretical energy and specific heat profiles are given for globular clusters.

## 1. Introduction

Globular clusters (GCs) are stellar systems with masses in the range $10^4 \div 10^6 M_\odot$ and about $10^5$ stars. The core radius $r_c$, i.e the radial coordinate at which the surface brightness becomes one half of the central value, may reach 10 $pc$, whereas the tidal radius $r_t$, i.e. the size of the cluster allowed by the external galactic tidal field, is typically about $50pc$. Furthermore GCs are nearly spherical systems.

The observations of GCs luminosity profiles [1] show similar curves, depending only on different values of the star concentration, and so they can be fitted by the same empirical law, suggesting a unique distribution function (DF) for all the clusters. This function corresponds approximately to a Boltzmann distribution function (BDF) minus a constant, resulting by the effects of the host galaxy tidal forces, which make GCs be spatially limited and their density profiles vanish in $r_t$. In this scenario stars going beyond the tidal radius, imposed by the galaxy potential, are lost from the system and evaporate, while, in terms of statistical mechanics, this limit turns to a restriction of the phase space, which is no longer entirely available because stars with velocities larger than a limiting one evaporate from the cluster. On the other hand, stars have the possibility of increasing their velocity due to stellar encounters, important in GCs since the average binary relaxation time is shorter than their absolute age ($10 \div 13$ $Gyr$). Although the stars evaporation, the velocity distribution remains formally unchanged during the evolution of a GC, varying only the concentration of the system or, equivalently, the dimensionless central gravitational potential $W_0$. Therefore the system will evolve through a series of King models [3], every of which is described by the same DF with different values of $W_0$, which increases during GCs lifetime. Furthermore the evolution of GCs can be studied as a classical single mass King model in relation to thermodynamic instability phenomena. In fact stellar encounters strongly contribute to phase space mixing of stellar orbits and so thermodynamics plays a central role in the gravitational equilibrium and stability of these clusters.

Analytically the evolution of GCs is well described by the Fokker-Planck equation, where the nature of collisions in GCs are taken into account, of which the King distribution function (KDF) is a solution [2]. This equation can determine the characteristics of the DF relevant for the equilibrium configurations, whereas the galactic gravitational tidal effects are responsible for the confinement of the cluster. All these effects make the evolution of GCs be described as a series of thermodynamic transformations, where the functional form of the KDF is conserved, while the structural and thermodynamic parameters of the system evolve [6]. Therefore two competitive phenomena determine the evolution of GCs: *i*) stellar encounters, which tend to refresh the high velocities tail in the DF, and *ii*) evaporation of stars, which prevents its formation. This scenario lasts until the so-called gravothermal catastrophe [4], due to the presence of negative specific heat regions caused by the combined effects of gravity and thermodynamics.

In this scenario it is possible to define thermodynamic quantities in analogy with the classical Boltzmann gas, paying attention to the fact that, while a Boltzmann gas is in thermodynamic equilibrium, GCs are in a steady-state equilibrium. Therefore the KDF can be treated as a Boltzmann-like distribution, with an additive term $\phi$ in the Hamiltonian, called effective potential. This potential is nothing else but the effect of the host galaxy tidal forces and causes the restriction of the phase space available to stars.

In this paper a generalized form of thermodynamics is obtained due to the introduction of the effective potential in the Hamiltonian. In section 2 the main thermodynamic quantities are obtained through the standard definitions of thermodynamics, stressing the fact there is a splitting of variables due to the presence of the effective potential. In section 3 the thermal, mechanical and chemical equilibria are studied, focusing the attention on the meaning of the two sets of variables. Finally in section 4 the behaviour of the specific heat and of the energy of the system is described, giving the possible theoretical configurations of GCs according to the exposed model, which can lead to the gravothermal catastrophe.

## 2. Fundamental principles

The KDF [2], describing the energy distribution of an isotropic and spherical system of stars with the same mass $m$, can be written as

$$f(\varepsilon) = \begin{cases} Ae^{-m\varphi/kT}\left[e^{-\varepsilon/kT} - e^{-\psi/kT}\right], & \varepsilon \leq \psi \\ 0, & \varepsilon > \psi \end{cases} \tag{1}$$

where $\psi(r) = m(\varphi_R - \varphi(r))$ is the energy cutoff imposed by the tidal interaction with the host galaxy, while $\varphi(r)$ is the gravitational potential. $A = Be^{(\mu+m\varphi)/kT}$, where $B$ is a constant of normalization and $\mu$ the chemical potential. Since its definition, $\psi$ is the energy sufficient to a star to reach the boundary of the system with zero velocity. Merafina & Ruffini studied GCs equilibrium configurations, showing they depend on the value of $W_0$ and the total mass of the system presents a maximum at $W_0 \simeq 1.35$, evidencing the rise of an instability [5]. Starting from first principles the evolution of GCs can be described as a series of thermodynamic transformations, where the parameters of the system evolve, since binary stellar encounters are important and evaporation takes place [6]. In this scenario thermodynamics is fundamental in the description of GCs and the above KDF can be rewritten as a Boltzmann-like DF, thanks to the introduction of an effective potential $\phi = \phi(\varepsilon, r)$

$$f(\varepsilon) = Ae^{-H/k\theta} \tag{2}$$

Journal of Physics: Conference Series **566** (2014) 012024 doi:10.1088/1742-6596/566/1/012024

where $A = Be^{(\alpha+m\varphi)/k\theta}$, $H = \varepsilon + m\varphi + \phi$ is the total Hamiltonian and $\theta$ the thermodynamic temperature. $B$ is a normalization constant, $\alpha$ the thermodynamic chemical potential and $\phi$ describes the effects of the host galaxy tidal forces. In the case of KDF

$$\phi = -k\theta \ln\left[1 - e^{(\varepsilon-\psi)/k\theta}\right] . \tag{3}$$

The introduction of an additive potential, which goes to infinity at $\psi$, leads to two different consequences, i.e. the restriction of the phase space and the recover of the BDF form. These effects lead to a natural splitting of thermodynamic variables, which have to be distinguished in thermodynamic quantities, which have the same characteristics of BDF variables and involve the full Hamiltonian, and kinetic quantities, which are the physical variables of the system and involve only the kinetic part of $H$. Using standard statistical expressions, the kinetic and thermodynamic variables are obtained

$$U_k = \bar{A}V \int_0^\psi f\ \varepsilon\ \varepsilon^{1/2}\ d\varepsilon \qquad U = \bar{A}V \int_0^\psi f\ H\ \varepsilon^{1/2}\ d\varepsilon \tag{4}$$

$$kT = \left\langle q_i \frac{\partial \varepsilon}{\partial q_i} \right\rangle \qquad k\theta = \left\langle q_i \frac{\partial H}{\partial q_i} \right\rangle \tag{5}$$

$$P = \frac{1}{3}\bar{A} \int_0^\psi f\ q\ \frac{d\varepsilon}{dq}\ d^3q \qquad \Pi = \frac{1}{3}\bar{A} \int_0^\psi f\ q\ \frac{dH}{dq}\ d^3q \tag{6}$$

$$\mu = \left.\frac{\partial U_k}{\partial N}\right|_{S,V} \qquad \alpha = \left.\frac{\partial U}{\partial N}\right|_{S,V} \tag{7}$$

where $q_i$ are the generalized momenta and $\bar{A}$ is a constant, while the brackets indicate the average over the DF. For number of particles and entropy

$$N = \bar{A}V \int_0^\psi f\ \varepsilon^{1/2}\ d\varepsilon \tag{8}$$

$$S = \bar{A}Vk \int_0^\psi (1 - \ln f) f \varepsilon^{1/2} d\varepsilon . \tag{9}$$

There is also a splitting of the first principle, Gibbs-Duhen relation, Euler equation and equation of state

$$dU_k = TdS - PdV + \langle\mu_0\rangle dN + N\langle d\nu\rangle \qquad dU = \theta dS - \Pi dV + \alpha dN + N\langle d\phi\rangle \tag{10}$$

$$N\langle d\nu\rangle = SdT - Vdp + Nd\langle\mu_0\rangle \qquad N\langle d\phi\rangle = Sd\theta - Vd\Pi + Nd\alpha \tag{11}$$

$$U_k = TS - PV + N\langle\mu_0\rangle \qquad U = \theta S - \Pi V + N\alpha \tag{12}$$

$$PV = NkT \qquad \Pi V = Nk\theta \tag{13}$$

where $\langle d\nu\rangle = d\langle\mu_0\rangle - \langle d\mu_0\rangle$. So the introduction of $\phi$ leads to a natural splitting not only of the thermodynamic variables, but also of the fundamental equations of thermodynamics.

## 3. Thermal, mechanical and chemical equilibria

Considering the results obtained in the previous section, it is possible to study the thermal, mechanical and chemical equilibria starting from the first principle. First of all two different types of variations have to be taken into account. The first type, labelled by $d$, is along a thermodynamic transformation, while the second one, labelled by $\delta$, is along the radial coordinate since quantities depend on $r$. So it is possible to study two infinitesimal near shells, at distance $\delta r$ one from each other, among which there is a difference of gravitational potential $\delta\phi$, divided by a separator, which makes the two parts not interact thermally, neither mechanically nor chemically between themselves. Now considering the first principle is true in every shell separately and the total entropy has its maximum value when the equilibrium is established ($dS_{tot} = 0$), for thermal, mechanical and chemical equilibria it will be

$$N,V = const \rightarrow \qquad \delta\theta = 0, \qquad k\delta T = R\left<\delta\phi\right> - (1-R)\, m\delta\phi \tag{14}$$

$$S,N = const \rightarrow \qquad \delta\Pi + \frac{N}{V}\left[\left<\delta\phi\right> + m\delta\varphi\right] = 0, \qquad \delta P + \rho\delta\varphi = 0 \tag{15}$$

$$V,S = const \rightarrow \qquad \delta\alpha = 0, \qquad N\delta\mu = -S\delta T \tag{16}$$

where $R = R(r)$ relates thermodynamic and kinetic variables and is given by

$$R = 1 - \frac{1}{3k\theta}\left< q\frac{\partial\phi}{\partial q}\right> \tag{17}$$

As it was revealed before, thermodynamic quantities have the same characteristics of BDF variables. Actually the thermodynamic temperature and chemical potential are constant for all configurations, whereas the kinetic variables are not: the former are only mathematical variables, the latter have physical meaning and must have a decreasing profile which goes to zero at the border of the configuration due to the characteristics of KDF. However the most important thing happens for the kinetic pressure, where the hydrostatic equilibrium comes out naturally from the mechanical equilibrium.

## 4. Specific heat and energy profiles: the gravothermal catastrophe

Two fundamental quantities for the analysis of GCs stability are the specific heat and the energy, which have a profile in radius alike the case of BDF. For what regards the energy of the system, it is made up of three terms:

$$E_{tot} = E_k + E_g + E_{eff} \tag{18}$$

where the last term takes into account the contribution of $\phi$ and represents the effect of the tidal forces caused by the host galaxy potential, which imposes the confinement of the system. The specific heat at constant volume is given by $C_V = (dQ/d\theta)_V$ from classical thermodynamics, and from first principle for thermodynamic variables

$$dQ = dU - N\left<dH\right> \tag{19}$$

where in the second term there is the contribution due to the effective potential. The results from numerical integration are shown in figs. 1-4. From the numerical analysis four different GCs configurations, and their evolution, can be predicted according to different values of $W_0$:

- $W_0 < 1.35$ positive specific heat all over the configuration and total positive energy: systems can not evolve towards the gravothermal catastrophe and disrupt due to $E_{tot} > 0$;
- $1.35 < W_0 < 2.3$ positive and negative specific heat regions with total positive energy: systems may evolve towards the gravothermal catastrophe but disrupt due to $E_{tot} > 0$;
- $2.3 < W_0 < 3$ positive and negative specific heat and energy regions with total positive energy and total negative specific heat: systems may evolve towards the gravothermal catastrophe but disrupt due to $E_{tot} > 0$;
- $W_0 > 3$ positive and negative specific heat and energy regions with total negative energy and specific heat: systems can evolve towards the gravothermal catastrophe by increasing $W_0$.

Journal of Physics: Conference Series **566** (2014) 012024 doi:10.1088/1742-6596/566/1/012024

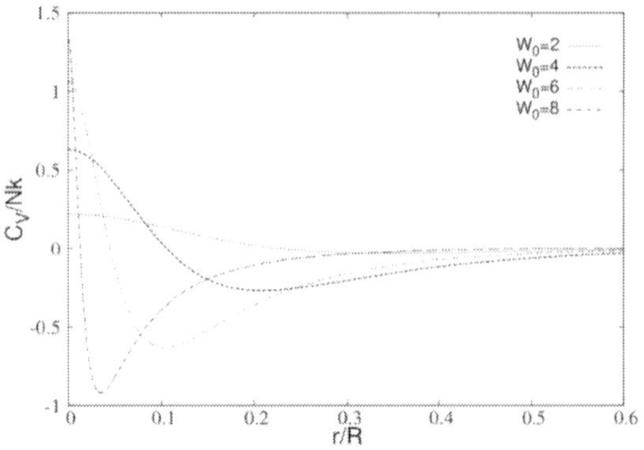


**Figure 1.** Radial profiles of the specific heat for different values of $W_0$.

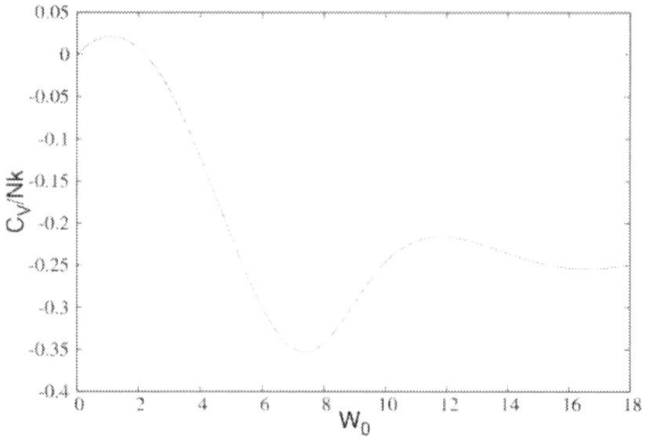


**Figure 2.** Total specific heat in function of $W_0$. Note it is negative for $W_0 > 2.3$.

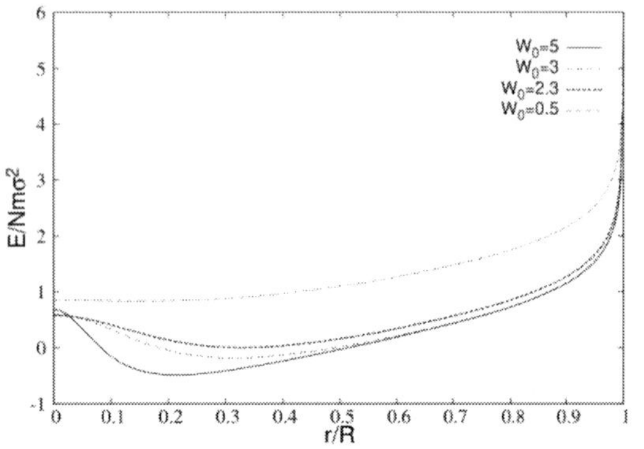


**Figure 3.** Radial profiles of the energy for different values of $W_0$.

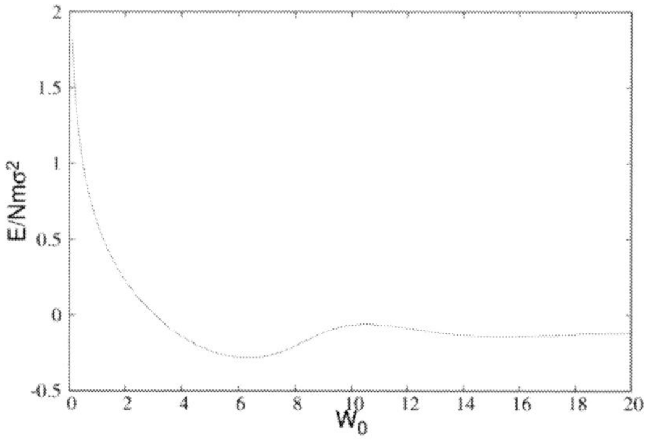


**Figure 4.** Total energy in function of $W_0$. Note it is negative for $W_0 > 3$.

## 5. Conclusions

The theoretical model exposed in this paper predicts negative and positive specific heat regions, so being self-consistent, for GCs, which can evolve towards the gravothermal catastrophe without an external bath of heat, alike the Lynden-Bell & Wood model [4]. The positive specific heat core could justify the presence of post core-collapsed objects. The system gives the value of $W_0 \simeq 6.9$ as the limit for the gravothermal instability and finds also preliminary observational evidences, by performing statistical analysis using Harris data [7][8].

## Acknowledgments

I want to thank the Organizers of the YRM 2014 for the possibility of publishing this proceeding and M.Merafina for his collaboration in the development of the exposed arguments.

## References

[1] King I R 1962 *AJ* **67** 471
[2] King I R 1965 *AJ* **70** 376
[3] King I R 1966 *AJ* **71** 64
[4] Lynden-Bell D and Wood R 1968 *MNRAS* **138** 495
[5] Merafina M and Ruffini R 1989 *A&A* **221** 4
[6] Horwitz G and Katz J 1977 *ApJ* **211** 226
[7] Harris W E 1996 *AJ* **112** 1487
[8] Merafina M and Vitantoni D 2014a *Acta Polyt.*

Journal of Physics: Conference Series **566** (2014) 012025 doi:10.1088/1742-6596/566/1/012025

# Modified gravity, the Cascading DGP model and its critical tension

**Fulvio Sbisà**[1,2]

[1] Institute of Cosmology and Gravitation, University of Portsmouth, UK

[2] Dipartimento di Fisica dell'Università di Milano and INFN, Sezione di Milano, Via Celoria 16, I-20133 Milano, Italy

E-mail: `fulvio.sbisa@gmail.com`

**Abstract.** We investigate the presence of instabilities in the Cascading DGP model. We start by discussing the problem of the cosmological late time acceleration, and we introduce the modified gravity approach. We then focus on brane induced gravity models and in particular on the Cascading DGP model. We consider configurations of the latter model where the source term is given simply by vacuum energy (pure tension), and we study perturbations at first order around these configurations. We perform a four-dimensional scalar-vector-tensor decomposition of the perturbations, and show that, regarding the scalar sector, the dynamics in a suitable limit can be described by a master equation. This master equation contains an energy scale (critical tension) which is related in a non-trivial way to the parameters of the model. We give a geometrical interpretation of why this scale emerges, and explain its relevance for the presence of ghost instabilities in the theory. We comment on the difference between our result and the one present in the literature, and stress its importance regarding the phenomenological viability of the model. We finally provide a numerical check which confirms the validity of our analysis.

## 1. Introduction

More than 15 years ago, cosmologists had a satisfying picture of how the universe worked, based on three main hypotheses: it was assumed that the universe is homogeneous and isotropic on large scales, that energy and momentum exist in the form of "ordinary" matter and radiation plus Cold Dark Matter (CDM), and that gravity is described by Einstein's General Relativity (GR). This Standard Cosmological model was successful in explaining several aspects of our universe, and predicted that the expansion of the universe necessarily has to decelerate. In fact, by the second Friedmann equation, the scale factor of the Universe $a(t)$ obeys $\ddot{a}/a \propto -(\rho + 3p)$ (where $\rho$ is the energy density, $p$ is the pressure and a dot indicates a time derivative), and the right hand side of the previous relation is negative for ordinary matter, radiation and CDM. The discovery of the present acceleration of the expansion [1, 2], successively confirmed by a variety of observations, forces us to abandon some of the assumptions mentioned above. We could relax the assumption of homogeneity, which differently from isotropy is not well-tested, and consider Lemaître-Tolman-Bondi models where the Earth is placed inside a large void. Another possibility is that the backreaction of the deviations from homogeneity and isotropy on the evolution of the scale factor is not negligible, and neglecting this effect may cause a misinterpretation of the observational data. If instead we enforce homogeneity, we could postulate that the universe is filled by a source (which we don't observe in the lab or in the solar system)

Journal of Physics: Conference Series **566** (2014) 012025 doi:10.1088/1742-6596/566/1/012025

which has negative pressure: the simplest option is the Cosmological Constant ($\Lambda$), but we may also consider new dynamical degrees of freedom (the *dark energy* paradigm). Finally, it may be that the cosmological observations are simply signalling that General Relativity is not the correct description of gravity at very large scales (the *modified gravity* approach). See [3, 4] and references therein for a more detailed discussion of these approaches.

The most straightforward solution of this problem seems to allow for a non-zero cosmological constant (*$\Lambda$CDM models*). However, fitting the observational data gives a value for $\Lambda$ which is really puzzling. If $\Lambda$ is considered as a new scale for gravity, then the universe is very fine tuned (coincidence problem), while if it is considered as the semi-classical manifestation of vacuum energy then the mismatch with the theoretical prediction is dramatic ("old" cosmological constant problem) [5]. Therefore, it seems worthwhile to pursue alternative approaches, and we choose to investigate the modified gravity approach.

## 2. Modified gravity and Cascading DGP

Modifying gravity is however potentially dangerous, since quite in general it introduces extra degrees of freedom compared to GR. On the one hand, these extra degrees of freedom have to be screened on terrestrial and astrophysical scales, otherwise they would be detected via fifth-force experiments and precise tests of gravity in the solar system. On the other hand, when studying the perturbative stability of certain solutions of the equations of motion, quite often some perturbation modes turn out to have negative kinetic energy and therefore lead to ghost instabilities. In fact, the presence of fields with negative kinetic energy (ghosts) renders the Hamiltonian unbounded from below, and causes the system to be unstable with respect to the simultaneous excitation of ghost and non-ghost fields (see [6] and references therein). Several modified gravity theories have been proposed so far, including $f(R)$ gravity, massive gravity, and braneworlds (see [7] for a review). In particular, braneworld models are appealing from a fundamental point of view, since the presence of extra dimensions and branes is crucial in string theory. In these theories, the spacetime has more than four dimensions and matter and radiation (as well as the strong, weak and electro-magnetic interactions) are confined on structures (branes) whose dimensionality is lower than the dimensionality of the ambient spacetime (bulk). Only gravity can propagate in the extra dimensions, which is consistent with the fact that in string theory gravity is described by closed strings. The codimension of a brane is defined as the difference between the dimension of the bulk and the dimension of the brane, and the embedding functions are those functions which indicate the position in the bulk of the points belonging to the brane. In general the confinement is not sharp, so matter, radiation and the gauge fields are distributed around the brane within a characteristic distance, which is determined by the properties of the system and of the confinement mechanism. This characteristic length scale is called the thickness of the brane, while the details of the distribution constitute the internal structure of the brane.

Concerning the late time acceleration problem, a promising idea is that of brane induced gravity, whose first and best-known realization is the DGP model [8]. Braneworld theories with induced gravity are characterized by the inclusion, in the part of the action which describes the dynamics of the brane, of an Einstein-Hilbert term built from the metric induced on the brane. This term can be introduced at classical level purely on phenomenological grounds, but can be also understood as a contribution coming from loop corrections in the low energy effective action of a quantum description where matter is confined on the brane [8]. The DGP model has the intriguing property of admitting self-accelerating cosmological solutions, which open the possibility of explaining the cosmic acceleration by geometrical means. On the other hand, it has been shown that these solutions have a perturbative ghost instability, and that the DGP model fits the cosmological data significantly worse than $\Lambda$CDM. See [9] and references therein for a discussion on the DGP model and its problems. A natural idea to go beyond the DGP model is to increase the codimension, while still having infinite extra dimensions and brane induced gravity. Such models offer also the possibility of addressing the Cosmological Constant problem, since they evade Weinberg's no-go theorem [10, 11] and may realize the degravitation mechanism [12]. However, it is not clear if increasing the codimension helps with the ghost problem, since there are contradicting results [13, 14, 15]. Moreover, the gravitational field on a brane of codimension higher than one diverges when the thickness of the brane tends to zero [16, 17, 18], and it is necessary to take explicitly into account its internal structure.

### *2.1. The Cascading DGP model*
Both of the aforementioned problems were claimed to be solved in a fairly recent class of models, the Cascading DGP [19]. In these models there is a recursive embedding of branes into branes of increasing dimensionality, each equipped with an appropriate induced gravity term. In the minimal set-up, a four-dimensional (4D) brane (our universe) is embedded inside a 5D brane which in turn is embedded in the 6D bulk. This model has three parameters, the masses $M_6$, $M_5$ and $M_4$, where $M_6^4$ controls the strength of the bulk action, $M_5^3$ controls the strength of the induced gravity term on the 5D brane and $M_4^2$ controls the strength of the induced gravity term on the 4D brane. The requirement that the model reproduces Einstein gravity on small scales fixes $M_4$ to be equal to the Planck mass, so this model has in truth two free parameters, the mass scales $m_6 = M_6^4/M_5^3$ and $m_5 = M_5^3/M_4^2$. They respectively control the relative strength between the bulk action and the induced gravity term on the 5D brane ($m_6$), and between the induced gravity term on the 5D brane and the induced gravity term on the 4D brane ($m_5$). It was shown that, moving from large distances to small distances, weak gravity "cascades" $6D \to 5D \to 4D$ when $m_6 \ll m_5$ while there is a direct transition $6D \to 4D$ when $m_6 \gg m_5$. Moreover, it was claimed that the presence of the codimension-1 brane with induced gravity renders the gravitational field finite on the codimension-2 brane even when the thickness of the latter tends to zero [19, 20]. A very important class of configurations are those that correspond to a source term given simply by vacuum energy ($\bar{\lambda}$) on the 4D brane (pure tension solutions). Importantly, it was claimed that there exists a critical tension $\bar{\lambda}_c$ such that a pure tension configuration is free of ghost instabilities (at first order in perturbations) if $\bar{\lambda} > \bar{\lambda}_c$ while it has ghosts if $\bar{\lambda} < \bar{\lambda}_c$ [19, 21].

## 3. The nested brane realization of the Cascading DGP model
To probe the dynamics of the internal structure of a brane, we need to excite it with amounts of energy roughly of the order of the inverse of the thickness (in "natural units" $\hbar = c = 1$). If we are interested only in what happens outside of the brane, and want to focus on energy scales lower than the inverse of the thickness, it is usual to consider a "thin limit description" in which the thickness of the brane is sent to zero while keeping fixed the amount of energy and momentum on the brane. In this case the brane is said to be "thin". While this is extremely useful for codimension-1 branes, it was proved that the thin limit of branes of codimension higher than one is not well-defined [22]. This result is very likely to be true also for more elaborate constructions such as in the Cascading case. Therefore, to obtain a well-defined formulation of the Cascading DGP model, it is necessary to give some information on the internal structure of the branes.

If we assume that there is a hierarchy between the thickness of the branes, namely that the codimension-1 brane is much thinner than the codimension-2 brane, we can describe the system as if a "ribbon" codimension-2 brane was present inside a thin codimension-1 brane. Furthermore, it can be shown that the thin limit of the ribbon brane (inside the already thin codimension-1 brane) is well defined [23]. It follows that, with this assumption about the hierarchy of the thicknesses, we can indeed work with thin branes and forget the internal structures. We refer to these set-ups as the *nested branes realization* of the Cascading DGP.

Journal of Physics: Conference Series **566** (2014) 012025 doi:10.1088/1742-6596/566/1/012025

### 3.1. *Perturbations around pure tension solutions*

To investigate the phenomenon which gives rise to the critical tension, we study perturbations at first order around solutions where pure tension is placed on the codimension-2 brane. These background solutions are most naturally expressed in a bulk-based approach, where the bulk metric is flat and the codimension-1 embedding has a cusp at the codimension-2 brane [23]. The complete space time is the product of a 4D Minkowski space and a 2D Riemannian cone, whose deficit angle is proportional to the tension. Therefore there exists a maximum tension $\bar{\lambda}_M = 2\pi M_6^4$ which corresponds to a deficit angle of $2\pi$ (when the cone becomes degenerate).

To study perturbations, we leave both the bulk metric and the codimension-1 embedding free to fluctuate. To deal with the issue of gauge invariance, we consider a 4D scalar-vector-tensor decomposition, and work with gauge-invariant variables [24]. In particular, in the scalar sector it is possible to express the equations in terms of two master variables, the trace part $\pi$ of the bulk metric perturbations, and the normal component of the codimension-1 bending $\delta\varphi_\perp$ [23] (we call *bending modes* the perturbation of the embedding functions). Notably, if we focus on the behaviour of the fields near the codimension-2 brane, it is possible to eliminate $\delta\varphi_\perp$ and obtain a master equation for $\pi$ alone. This equation however contains the derivative of $\pi$ normally to the codimension-1 brane, so to know the behaviour of $\pi$ on the codimension-2 brane it is still necessary to solve the full 6D problem. This difficulty can be overcome by considering a "4D limit", which gives the following (local) equation on the codimension-2 brane [25]

$$3M_4^2\left[1-\frac{3}{2}\frac{m_5}{m_6}\tan\left(\frac{\bar{\lambda}}{4M_6^4}\right)\right]\Box\pi=\mathcal{T} \tag{1}$$

where $\Box$ indicates the four-dimensional D'Alembert operator and $\mathcal{T}$ is the trace of the energy-momentum tensor. This equation indicates that $\pi$ is an effective 4D ghost if $0<\bar{\lambda}<\bar{\lambda}_c$ while it is a healthy field if $\bar{\lambda}_c<\bar{\lambda}<\bar{\lambda}_M$, where the *critical tension* reads $\bar{\lambda}_c \equiv 4M_6^4 \arctan(2m_6/3m_5)$.

## 4. Geometrical interpretation and ghost-free regions in parameters space

Having identified the critical tension, we can study the coupled dynamics of the fields $\pi$ and $\delta\varphi_\perp$ to interpret geometrically its existence. Considering the 4D limit mentioned above, it can be shown that the trace of the energy-momentum tensor $\mathcal{T}$ excites $\pi$ via two separate channels. It does so directly, because of the 4D induced gravity term, and indirectly via the bending of the codimension-1 brane, because of the 5D induced gravity term. Crucially, the first channel excites $\pi$ in a ghostly and $\bar{\lambda}$-independent way, while the second channel excites $\pi$ in a healthy and $\bar{\lambda}$-dependent way [25]. The existence of the critical tension is due to the competition between these two channels, and the fact that the field $\pi$ is a ghost or not is decided by the first channel being more or less efficient than the second channel. Note that the existence of the second channel is entirely due to the higher dimensional structure of the theory, and in particular to the presence of the induced gravity term on the codimension-1 brane.

Our result for the critical tension is at odds with the findings of [19, 21], which found the value $\bar{\lambda}_c^{dRKT} = 8\,m_6^2\,M_4^2/3$. These two results coincide when $m_6 \ll m_5$ but differ significantly when $m_6 \gtrsim m_5$ and are dramatically different when $m_6 \gg m_5$. To see why this difference is crucial, note that the value we found for the critical tension is always smaller than the maximum tension $\bar{\lambda}_M$, so for every value of $m_5$ and $m_6$ we can find a range of values for the background tension such that $\pi$ is a healthy field. However, $\bar{\lambda}_c^{dRKT}$ is smaller than $\bar{\lambda}_M$ only when $m_6 \lesssim m_5$, so the results of [19, 21] imply that half of the phase space of the theory is plagued by ghosts, and so is phenomenologically ruled out. It is therefore very important to establish why two different results are obtained, and which of the two is correct.

### *4.1. Numerical check*

A priori we may wonder if, referring to the discussion in section 3, the hypotheses used in [19, 21] about the internal structure of the branes are different from the one we use, and therefore in truth we are considering different models. However this is not the case, since our analysis can be mapped into that of [21] by a coordinate transformation [25]. In fact it can be shown that the difference lies in how the junction conditions at the codimension-2 brane are derived, which is linked to which hypotheses are made on the behaviour of the fields near the codimension-2 brane in the thin limit. Roughly speaking, we assume that the embedding functions remain continuous in the thin limit ("the codimension-1 brane does not break"), while the result of [21] is reproduced assuming that the normal component of the bending remains continuous. These two conditions cannot be both satisfied at the same time, since in the background solutions the normal vector becomes discontinuous in the thin limit [25]. To decide which of the two results is correct, we consider a case, the pure tension perturbation case, where the internal structure of the codimension-2 brane is exactly solvable. In this case it is possible to derive the codimension-2 junction conditions by performing numerically the integration of the codimension-1 junction conditions across the codimension-2 brane (pillbox integration). Since the configuration of the fields is known explicitly also inside the codimension-2 brane, it is not necessary to make any hypothesis on the behaviour of the fields to do that.

The idea is to perform the numerical pillbox integration for a sequence of configurations with different thickness, and study the asymptotic behaviour of the outcome when the thickness tends to zero. If our analysis or the analysis of [19, 21] is correct, then it has to agree with the numerical outcome in the thin limit. The outcome of the numerical integrations are plotted in figure 1 (the parameter $n$ is inversely proportional to the thickness). It is evident that the points corresponding to our codimension-2 junction conditions (squares) converge to the points corresponding to the numerical integration (circles), while the points corresponding to the codimension-2 junction conditions which reproduce the result of [21] (diamonds) are significantly distant from the former ones.

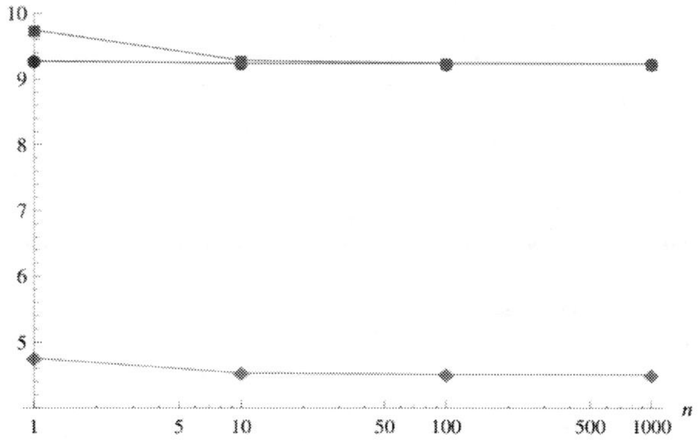


**Figure 1.** Plot of the numerical results of the pillbox integration

## 5. Conclusions

The numerical check above puts on firm footing our analysis of the nested branes realization of the 6D Cascading DGP, and strongly supports our claim that the correct value of the critical tension is $\bar{\lambda}_c \equiv 4M_6^4 \arctan\left(2m_6/3m_5\right)$. In particular, it strongly suggests that also models where gravity displays a direct transition 6D $\rightarrow$ 4D ($m_6 > m_5$) are phenomenologically viable. We conclude that braneworld models with infinite volume extra dimensions and induced gravity might be a powerful tool to tackle the late time acceleration and the Cosmological Constant problems. In particular, the Cascading DGP seems a promising candidate to overcome the problems of the DGP models whilst preserving its good features. On the other hand, the singular structure of the geometry is very subtle at the codimension-2 brane, thus indirectly confirming the belief that the singular structure of branes of codimension higher than one is in general more complex than that of codimension-1 branes.

Regarding future directions of research, a lot of work is still to be done, including studying the existence of the ghost at full non-linear level by performing a Hamiltonian analysis. More in general, it is important to derive the codimension-2 junction conditions at non-perturbative level and to derive cosmological solutions, as well as verifying if the model passes the high-precision solar system tests.

## Acknowledgments

This talk is based on work done in collaboration with Kazuya Koyama.

## References

[1] Riess A G *et al.* (Supernova Search Team) 1998 *Astron. J.* **116** 1009–38 (*Preprint* astro-ph/9805201)
[2] Perlmutter S *et al.* (Supernova Cosmology Project) 1999 *Astrophys. J.* **517** 565–86 (*Preprint* astro-ph/9812133)
[3] Ellis G, Nicolai H, Durrer R and Maartens R 2008 *Gen. Rel. Grav.* **40** 219–20
[4] Durrer R and Maartens R 2008 *Dark Energy: Observational & Theoretical Approaches* ed. P Ruiz-Lapuente (Cambridge UP, 2010) 48–91 (*Preprint* 0811.4132)
[5] Weinberg S 1989 *Rev. Mod. Phys.* **61** 1–22
[6] Sbisa' F 2014 Classical and quantum ghosts *Preprint* 1406.4550
[7] Clifton T, Ferreira P G, Padilla A and Skordis C 2012 *Phys. Rept.* **513** 1–189 (*Preprint* 1106.2476)
[8] Dvali G, Gabadadze G and Porrati M 2000 *Phys. Lett. B* **485** 208–214 (*Preprint* hep-th/0005016)
[9] Sbisa' F 2014 Modified theories of gravity *Preprint* 1406.3384
[10] Arkani-Hamed N, Dimopoulos S, Dvali G and Gabadadze G 2002 Nonlocal modification of gravity and the cosmological constant problem *Preprint* hep-th/0209227
[11] Dvali G, Gabadadze G and Shifman M 2003 *Phys. Rev. D* **67** 044020 (*Preprint* hep-th/0202174)
[12] Dvali G, Hofmann S and Khoury J 2007 *Phys. Rev. D* **76** 084006 (*Preprint* hep-th/0703027)
[13] Dubovsky S and Rubakov V 2003 *Phys. Rev. D* **67** 104014 (*Preprint* hep-th/0212222)
[14] Kolanovic M, Porrati M and Rombouts J W 2003 *Phys. Rev. D* **68** 064018 (*Preprint* hep-th/0304148)
[15] Berkhahn F, Hofmann S and Niedermann F 2012 *Phys. Rev. D* **86** 124022 (*Preprint* 1205.6801)
[16] Cline J M, Descheneau J, Giovannini M and Vinet J 2003 *J. High Energy Phys.* JHEP06(2003)048 (*Preprint* hep-th/0304147)
[17] Vinet J and Cline J M 2004 *Phys. Rev. D* **70** 083514 (*Preprint* hep-th/0406141)
[18] Bostock P, Gregory R, Navarro I and Santiago J 2004 *Phys. Rev. Lett.* **92** 221601 (*Preprint* hep-th/0311074)
[19] de Rham C, Dvali G, Hofmann S, Khoury J, Pujolas O *et al.* 2008 *Phys. Rev. Lett.* **100** 251603 (*Preprint* 0711.2072)
[20] de Rham C, Hofmann S, Khoury J and Tolley A J 2008 *J. Cosmol. Astropart. Phys.* JCAP02(2008)011 (*Preprint* 0712.2821)
[21] de Rham C, Khoury J and Tolley A J 2010 *Phys. Rev. D* **81** 124027 (*Preprint* 1002.1075)
[22] Geroch R and Traschen J 1987 *Phys. Rev. D* **36** 1017
[23] Sbisa' F and Koyama K 2014 *J. Cosmol. Astropart. Phys.* JCAP06(2014)029 (*Preprint* 1404.0712)
[24] Mukohyama S 2000 *Phys. Rev. D* **62** 084015
[25] Sbisa' F and Koyama K 2014 *J. Cosmol. Astropart. Phys.* JCAP09(2014)038 (*Preprint* 1405.7617)

# Light-cone effect and relaxation after a quantum quench from excited states in the Ising chain

**Leda Bucciantini**

Dipartimento di Fisica dell'Università di Pisa and INFN, 56127, Pisa, Italy

E-mail: `leda.bucciantini@df.unipi.it`

**Abstract.** The time evolution and stationary behaviour of local observables is studied after a quantum quench of the magnetic field in the transverse field Ising chain, when the initial state is chosen as an excited state of the pre-quench Hamiltonian. Relaxation to the Generalized Gibbs Ensemble is proved. Equal-time longitudinal two point function and entanglement entropy are analytically derived for a class of excited states which confirm the light-cone spreading of correlations also starting from excited states.

## 1. Introduction

Recent advances in the technology of ultra-cold atoms [1] have made it possible to experimentally realize quantum systems which represent nearly ideal testing grounds for the study of non-equilibrium isolated many-body quantum systems. This has led to the possibility of addressing long standing theoretical questions concerning equilibration and thermalisation [2], the role of integrability and universality [3] in out-of-equilibrium many-body quantum systems.

The simplest way of driving a system out of equilibrium is through a sudden quantum quench [4]; it consists in preparing a system in an initial pure state, that is an eigenstate of the pre-quench Hamiltonian $H$, then suddenly varying a parameter of the Hamiltonian and letting the system evolve unitarily, i.e. without any coupling to the external environment, according to a post-quench Hamiltonian $H'$, with $[H, H'] \neq 0$.

The two main results derived from both experimental and theoretical investigations are (a) the existence of a maximum value for the velocity of spreading of correlations in a lattice system considered in a purely quantum mechanical framework [5] and (b) the long-time relaxation of local observables to time independent values, when the thermodynamic limit is taken over the whole system.

Interestingly enough, the stationary values of the subsystem can be predicted using a statistical ensemble formulation, *as if* the density matrix of the whole system be in a mixed state and tracing out the degrees of freedom external to the subsystem [6]. The general picture is that non integrable systems thermalise [2], i.e. they can be described by the Gibbs or microcanonical ensemble which retains only very few information of the initial state, while integrable[1] ones relax to the Generalized Gibbs Ensemble (GGE) [7], which takes into account all local mutually commuting conserved charges. This scenario has been widely verified in numerical, experimental and theoretical analysis, although some exceptions have been found for some specific initial states

[1] For some insights in the definition of quantum integrability see for instance [8].

[9]; clearly, the most convincing evidence supporting it comes from the exactly solvable models, whose dynamics can be worked out explicitly.

The role of initial states has also proved to be crucial for the time evolution following a quantum quench [10] but I would like to point out that it has always been taken for granted that the initial state was the ground state of the pre-quench Hamiltonian. In this work I set up to thoroughly study, for the first time to my knowledge, stationary and dynamical behaviour of local observables after a quench in an exactly solvable model, starting with initial states that are highly excited state of the pre-quench Hamiltonian. Its interest should rely on the radically different behaviour of entanglement entropy for ground states of local Hamiltonians (following an area law, with at most multiplicative logarithmic corrections [11]) compared to that of highly excited states (extensive in subsystem's size [12]).

In the next sections I will present the model studied and show that in the long-time limit, taken after the thermodynamic one, the system can be locally described by the GGE; I will then address the time evolution of some relevant observables, namely the transverse magnetisation, the longitudinal two-point function and entanglement entropy, highlighting the differences obtained with respect to the ground-state case.

## 2. Model

In order to shed some light on this problem I will consider here, among the exactly solvable models, the transverse field Ising chain which, despite its simplicity, represents a crucial paradigm for quantum critical behaviour. Its Hamiltonian is

$$H(h) = -\frac{1}{2}\sum_{j=1}^{N}\left[\sigma_j^x\sigma_{j+1}^x + h\sigma_j^z\right], \tag{1}$$

where $\sigma_j^\alpha$, $\alpha = x, y, z$ are the Pauli matrices at site $j$ of the chain of length $N$, $h$ is the transverse field and periodic boundary conditions are imposed. The model can be mapped to spinless free fermions through the non-local Jordan–Wigner transformation and diagonalised by a further Bogoliubov transformation in momentum space, yielding [13]

$$H(h) = \sum_k \epsilon_k \left(b_k^\dagger b_k - \frac{1}{2}\right), \tag{2}$$

where $b_k$, $b_k^\dagger$ are the annihilation and creation operator for the fermionic quasi-particles of momentum $k$, satisfying $\{b_k, b_{k'}^\dagger\} = \delta_{kk'}$, with one-particle dispersion relation given by $\epsilon_k = \sqrt{(h-\cos p_k)^2 + \sin^2(p_k)}$, with $p_k = \frac{2\pi k}{N}$.

I will focus on quenches when at $t = 0$ the transverse magnetic field is suddenly switched from $h_0$ to $h$. As initial state I will consider an excited state of the pre-quench Hamiltonian $H(h_0)$ which is neither an eigenstate nor a finite superposition of eigenstates of $H(h)$ of the form[2]

$$|\Psi_0\rangle = \prod_k (b'^\dagger_k)^{m_k}|0\rangle, \tag{3}$$

where $m_k = 0, 1$ is the fermionic initial occupation number of the $k$-mode (i.e. the excitation profile characterizing the initial state $|\Psi_0\rangle$) while $|0\rangle$ is the ground state of the pre-quench Hamiltonian. In the thermodynamic limit, $m_k$ becomes an arbitrary function $m(\varphi) \in [0, 1]$, with $\varphi \in [-\pi, \pi]$ and represents a coarse-grained version of $m_k$ [12]. I would like to point out that Eq. (3) is not the most generic excited state, since I have focused only on states that are a basis for the many-body Hilbert space, and not linear superposition of them.

[2] From now on, primed quantities will be used to denote pre-quench operators, while non-primed quantities for post-quench ones.

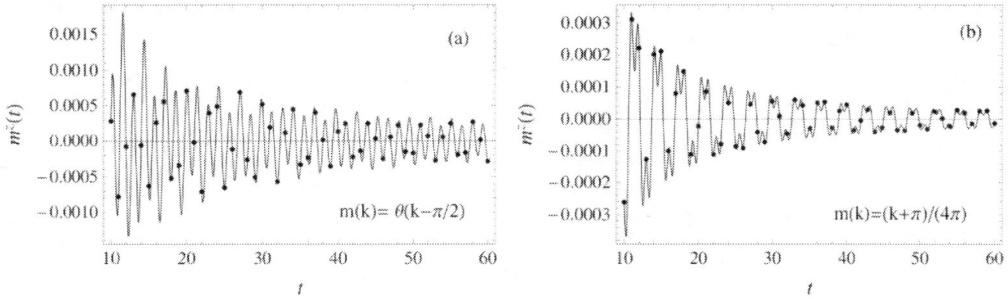


**Figure 1.** Transverse magnetisation minus its stationary value as a function of time for quenches from $h_0 = 12$ to $h = 2$ starting from (a) $m(k) = \theta(k - \pi/2)$ and (b) $m(k) = (k + \pi)/(4\pi)$.

## 3. Stationary limit

The computation of the reduced density matrix for a block of $\ell$ contiguous spins in free theories is a much simpler task than in generic integrable ones: while the latter has $2^\ell \times 2^\ell$ non trivial matrix-elements, the former can be written using Wick's theorem in terms of a $2\ell \times 2\ell$ block Toeplitz matrix, whose elements can be computed only from the fermionic two-point function [15]. Given this direct relation between reduced density matrix and two-point function correlation matrix, it is sufficient to prove that the stationary ensemble and GGE have the same correlation matrix in order to prove that they are equal. Hence, first I compute the infinite time limit of the fermionic two-point function; then I compute the same expectation value using the Generalized Gibbs Ensemble constructed with the conserved fermionic post-quench occupation number $n_k = b_k^\dagger b_k$ [14]

$$\rho_{\text{GGE}} = \frac{e^{-\sum_k \lambda_k n_k}}{Z_{\text{GGE}}}, \tag{4}$$

where $\lambda_k$ are fixed by matching the expectation values of the conserved occupation numbers with their values in the initial state, i.e. imposing $\langle \Psi_0 | b_k^\dagger b_k | \Psi_0 \rangle = \text{Tr}[\rho_{\text{GGE}} b_k^\dagger b_k]$, and $Z_{\text{GGE}} = \text{Tr}[e^{-\sum_k \lambda_k n_k}]$. It can be shown that the two expressions coincide; this proves the validity of GGE in the transverse field Ising chain also starting from initial excited states.

## 4. Dynamics

I have analysed the time dependence of the transverse magnetization, of the longitudinal two-point function between spins sitting at a fixed finite distance $\ell$ $\rho^{xx}(\ell, t) \equiv \langle \psi_0(t) | \sigma_n^x \sigma_{\ell+n}^x | \psi_0(t) \rangle$ and of the entanglement entropy.

*Transverse magnetisation*

The analytic expression for the transverse magnetisation $m^z(t) = \langle \psi_0(t) | \sigma_i^z | \psi_0(t) \rangle$, with $|\psi_0(t)\rangle = e^{iH(h)t} |\psi_0\rangle$, in the thermodynamic limit is [16]

$$m^z(t) = \int_{-\pi}^{\pi} \frac{dk}{4\pi} e^{i\theta_k} m_k^S [\cos \Delta_k - i \sin \Delta_k \cos(2\epsilon_k t)], \tag{5}$$

where $\theta_k$ and $\Delta_k$ are the angle of the Bogoliubov rotation that respectively diagonalises the Hamiltonian and connects pre-quench with post-quench variables

$$\tan \theta_k = \frac{\sin(2\pi k/N)}{\cos(2\pi k/N) - h} \qquad \cos \Delta_k = \frac{h h_0 - (h + h_0) \cos \varphi_k + 1}{\sqrt{1 + h^2 - 2h \cos(\varphi_k)} \sqrt{1 + h_0^2 - 2h_0 \cos(\varphi_k)}}; \tag{6}$$

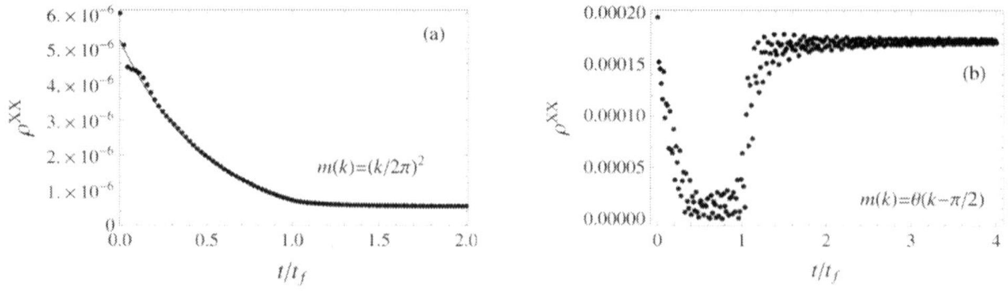


**Figure 2.** Time evolution of $\rho^{xx}(\ell, t)$ for quenches from $h_0 = 1/3$ to $h = 2/3$. It is reported in units of the Fermi time $t_F$, defined as $t_F = \frac{\ell}{2v_{\max}}$ where $v_{\max}$ is the maximal propagation velocity of the elementary excitations $v_{\max} = \max_{k \in [-\pi,\pi]} |\epsilon'_k| = \min[h, 1]$. Points represent the numerical evaluation by means of the determinant approach while the continuous line in (a) is the analytical prediction. (a) $m(k) = k^2/(2\pi)^2$ for $\ell = 60$; (b) $m(k) = \theta(k - \pi/2)$ for $\ell = 90$.

$m^S_k$ and $m^A_k$ are respectively the symmetrized and antisymmetrized part of $m(k)$, namely $m^S_k \equiv m_{-k} + m_k - 1$ and $m^A_k \equiv m_{-k} - m_k$.

The approach to stationary values can be evaluated by a stationary phase approximation and exhibits a different power-law decay depending on the analyticity of the $m(k)$ excitation profile. In Fig.1 I report the time dependent part of the transverse magnetisation for two different prototypical excited states (figures are taken from [16]): exact results obtained from the numerical evaluation of the integral in Eq. (5) (points) are compared with its stationary phase approximation (line), and it can be seen that the agreement is excellent. For an initial excitation profile given by $m(k) = \theta(k - \pi/2)$, Fig.1a, the power law decay goes like $t^{-1}$, while for $m(k) = (k + \pi)/(4\pi)$, Fig.1b, it goes like $t^{-\frac{3}{2}}$, as for the ground state. Thus, my conclusion is that while for quenches starting from the ground state the power law decay is $t^{-3/2}$, this is not true in general for excited states (see Fig.1b) which display a much slower relaxation $\propto t^{-1}$; actually the excited state for which that happens is the most physical one among those studied since it has all modes larger than a given one occupied with a fermion.

*Equal time two point longitudinal correlation function and entanglement entropy*

The time dependence of the two-point function $\rho^{xx}(\ell, t)$ has been computed numerically within the determinant approach for all initial excited states analysed [16]; I have also been able to obtain analytical results [16] in the scaling limit (i.e. $\ell \to \infty$, $t \to \infty$, with $\ell/t$ fixed) for quenches within the ferromagnetic phase and for initial excitation profiles preserving $k \to -k$ symmetry.

Calculations have been carried out for many excited states, with excitation profile $m(k)$ of linear, quadratic, or well shape. All the data obtained from these states show a quite general behaviour, exemplified in Fig. 2a: for $t < t_F$ the correlation function decays exponentially, while for $t > t_F$ it shows a slow relaxation towards the GGE value. This is a manifestation of the light-cone spreading of correlations [4] also for quenches from excited states. Still there is a fundamental difference with respect to the ground-state initial case: since the initial state is $Z_2$-invariant[3], the one point function $\langle \sigma^x_i \rangle$ is always vanishing, explaining why the initial value of the two-point functions in Fig. 2 is vanishing for increasing values of $\ell$, while it is close to 1 for the ground-state quench.

For one of the excited states, namely $m(k) = \theta(k - \pi/2)$, I found an unusual manifestation of the light-cone behaviour in the two-point function; as can be seen in Fig. 2b, after an initial decay

[3] This holds true apart from the two cases of $m(k) = 0$ or $m(k) = 1$ for any k.

the correlation function displays a sort of plateaux and at $t \sim t_F$ it sets around the GGE value which is reached in an oscillating manner. An intuitive explanation of this uncommon behaviour is presently lacking, even because the analytical method developed for $k \to -k$ symmetric states does not straightforwardly generalize to the state under consideration.

The entanglement entropy for a block of $\ell$ contiguous spins, defined as the Von Neumann entropy of the reduced density matrix of the subsystem, can be related to the eigenvalues of a $2\ell \times 2\ell$ block Toeplitz matrix constructed with the fermionic two point functions [17]. I evaluated its time dependent behaviour in a similar way as done for $\rho^{xx}(\ell, t)$; the result, common to all excited states analysed (for figures see [16]), shows a light-cone behaviour, i.e. a linear growth for $t < t_F$ followed by a slow saturation. The main difference with the ground state case is the fact that the entanglement entropy in the initial state is extensive.

## 5. Conclusions

I have considered the stationary and dynamical behaviour of several observables after a sudden quantum quench of the transverse magnetic field in the Ising chain, starting from an initial excited state. While still having relaxation towards the GGE, I have found some significant differences with respect to the initial ground state case, among which a different power law decay of the transverse magnetisation, and an unusual manifestation of the light-cone effect for a specific initial state. The role of excitations in truly interacting integrable models will be investigated in future works.

## References

[1] T. Kinoshita, T. Wenger, D. S. Weiss, Nature **440**, 900 (2006); S. Trotzky Y.-A. Chen, A. Flesch, I. P. McCulloch, U. Schollwöck, J. Eisert, and I. Bloch, Nature Phys. **8**, 325 (2012).

[2] J. M. Deutsch, Phys. Rev. A **43**, 2046 (1991); M. Srednicki, Phys. Rev. E **50**, 888 (1994).

[3] M. Rigol, V. Dunjko, and M. Olshanii, Nature **452**, 854 (2008); M. Olshanii, and V. Yurovsky, arXiv:0911.5587 (2009); C. De Grandi, V. Gritsev, and A. Polkovnikov, Phys. Rev. B **81**, 012303 (2010); A. Polkovnikov, K. Sengupta, A. Silva, and M. Vengalattore, Rev. Mod. Phys. **83**, 863 (2011).

[4] P. Calabrese and J. Cardy, Phys. Rev. Lett. **96**, 136801 (2006); P. Calabrese and J. Cardy, J. Stat. Mech. P06008 (2007).

[5] E. H. Lieb and D. W. Robinson, Commun. Math. Phys., **28**, 251 (1972). M. Cheneau, P. Barmettler, D. Poletti, M. Endres, P. Schauss, T. Fukuhara, C. Gross, I. Bloch, C. Kollath, and S. Kuhr, Nature **481**, 484 (2012); A. Laeuchli and C. Kollath, J. Stat. Mech. P05018 (2008); S. R. Manmana, S. Wessel, R. M. Noack, and A. Muramatsu, Phys. Rev. B **79**, 155104 (2009); P. Barmettler, D. Poletti, M. Cheneau, and C. Kollath, Phys. Rev. A **85**, 053625 (2012).

[6] P. Calabrese, F.H.L. Essler, and M. Fagotti, J. Stat. Mech. P07022 (2012). T. Barthel and U. Schollwöck, Phys. Rev. Lett. **100**, 100601 (2008).

[7] M. Rigol, V. Dunjko, V. Yurovsky, and M. Olshanii, Phys. Rev. Lett. **98**, 50405 (2007).

[8] J.-S. Caux, J. Mossel, J. Stat. Mech. P02023 (2011).

[9] B. Wouters, M. Brockmann, J. De Nardis, D. Fioretto and J.-S. Caux, arXiv:1405.0172; B. Pozsgay, M. Mestyan, M. A. Werner, M. Kormos, G. Zarand and G. Takacs, arXiv:1405.2843; G. Goldstein and N. Andrei, arXiv:1405.4224.

[10] S. Sotiriadis, G. Takacs and G. Mussardo1, Phys. Lett. B **734** (2014); M. C. Bauls, J. I. Cirac, and M. B. Hastings, Phys. Rev. Lett. **106**, 050405 (2011); C. Gogolin, M. P. Mller and J. Eisert, Phys. Rev. Lett. **106**, 040401 (2011).

[11] J. Eisert, M. Cramer, and M. B. Plenio, Rev. Mod. Phys. **82**, 277 (2010).

[12] V. Alba, M. Fagotti, and P. Calabrese, J. Stat. Mech. P10020 (2009).

[13] S. Sachdev, Quantum Phase Transitions, Cambridge University Press, 2001.

[14] M. Fagotti and F.H.L. Essler, Phys. Rev. B **87**, 245107 (2013)

[15] P. Calabrese, F.H.L. Essler, and M. Fagotti, J. Stat. Mech. P07016 (2012). P. Calabrese, F.H.L. Essler, and M. Fagotti, J. Stat. Mech. P07022 (2012).

[16] L. Bucciantini, M. Kormos, and P. Calabrese, J. Phys. A **47**, 175002 (2014)

[17] M. Fagotti and P. Calabrese, J. Stat. Mech. P04016 (2010).

# One–neutron transfer reaction: a toy model in one dimension

**L Moschini**

Dipartimento di Fisica e Astronomia "G. Galilei", Padova, Italy
INFN, Sezione di Padova, Padova, Italy

E-mail: `laura.moschini@pd.infn.it`

**Abstract.** A simple 1D toy model to study one–neutron transfer reactions is developed. It is based on the solution of the time dependent Schroedinger equation for a particle initially bound by a fixed potential well, perturbed by a second moving potential, which accounts for the second partner of the reaction. At the end of the time evolution it is possible to evaluate the probability of the transfer of the particle from a potential to the other, as well as the transfer to continuum states in the case of weakly–bound systems. Although rather simple, the model accounts for most of the physical characteristics of these kind of reactions: such as the existence of an optimum Q–value and the dependence on the parameters defining the relative motion of the two potentials.

## 1. Introduction

Nuclear reactions between heavy ions are traditionally and usefully divided in two groups, direct and compound reactions, depending on the parameters of the reaction, such as bombarding energy and impact parameter. Direct reactions go directly from initial to final state in a single step, while in compound reactions the projectile experiences subsequent collisions leading to the formation of an intermediate system. Of course, it may be possible that direct and compound-nucleus processes both contribute to a given reaction, but in general the greater the incident energy in a reaction, the less likely the possibility of multi step processes.

The usual characteristic of a direct reaction found in the experiments is a tendency for only small amounts of momentum to be transferred, thus the outgoing particle will carry most of the incident energy and momentum of the projectile; since the interaction is restricted mainly in the nuclear surface (peripheral collisions), a tipical signature is an angular distribution peaked around the grazing angle. Moreover direct processes occur very rapidly in a time of the order of $10^{-22}$ s, while a compound process typically takes much longer, perhaps $10^{-16}$ to $10^{-18}$ seconds. Among direct reaction there are different channels, each one characterized by a particular behavior: inelastic scattering excites collective states; one–neutron transfer probes the single–particle character of states; two–nucleon transfer goes preferentially to states that exhibit strong pairing correlations. It is possible that two–neutron, two–proton transfers may reveal alpha clustering, and so on. Since direct reactions are highly selective in the nuclear levels that they populate, they are useful tools to investigate the nuclear structure.

Transfer reactions are special kind of direct reaction in which one or more nucleons are transferred between the projectile and the target. Some example with light ions could be (d,p) and (t,p) reactions, extensively studied at radioactive beam facilities as ISOLDE, MSU, GANIL and TRIUMF. A scheme of a transfer reaction is presented in figure 1.

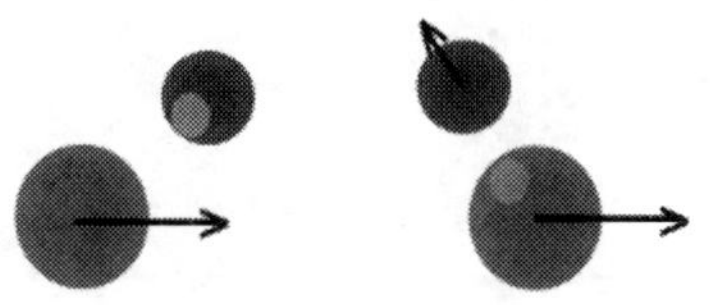

Figure 1: A transfer reaction scheme: an incoming nucleus picks up a cluster or a particle from a target nucleus.

Here I am particularly interested in describing one–neutron transfer reactions. To study this kind of reactions I am developing a 1D toy model. At variance with the usual couple–channel approach, which may become rather cumbersome and suffers from the inevitable truncation of the basis, I can provide an exact calculation for the time evolution of the system. A comparison between these two methods is presented in [1].

## 2. The model

I solve[1] the one–particle time dependent Schroedinger equation

$$i\hbar\frac{d}{dt}\Psi(x,t) = \mathcal{H}(x,t)\Psi(x,t) \tag{1}$$

with Hamiltonian

$$\mathcal{H}(x,t) = \frac{\hbar^2}{2\mu}\frac{d^2}{dx^2} + V_T(x) + V_P(x_P(t)). \tag{2}$$

The target mean–field potential is parametrized by a Woods–Saxon well

$$V_T(x) = \frac{V_T}{1+e^{\frac{|x|-R_T}{\alpha_T}}}. \tag{3}$$

The initial wave function $\Psi(x,t=0)$ is a single particle state in the target potential, calculated applying a Numerov algorithm [2]. The projectile potential

$$V_P(x) = \frac{V_P}{1+e^{\frac{|x-x_P(t)|-R_P}{\alpha_P}}} \tag{4}$$

moves with the time dependence of

$$x_P(t) = x_0 + \frac{1}{2}at^2, \tag{5}$$

where $x_0$ is the distance of the closest approach (the distance between the potentials centers). The calculation starts from a distance long enough to avoid interactions with the target.
The acceleration could be fixed taking into account $\mu a = -\frac{d}{dr}V_c(r)$ considering the reduced mass of the system, but, since I am interested to the response of the system at the variation of each parameter, I will consider it as an adjustable parameter.
In figure 2 some steps of the time evolution are reported. The upper part of each picture gives the square of the wave function, while the lower one gives the actual position of the two wells.

[1] I made use of D02PVF and D02PCF functions provided by NAG library.

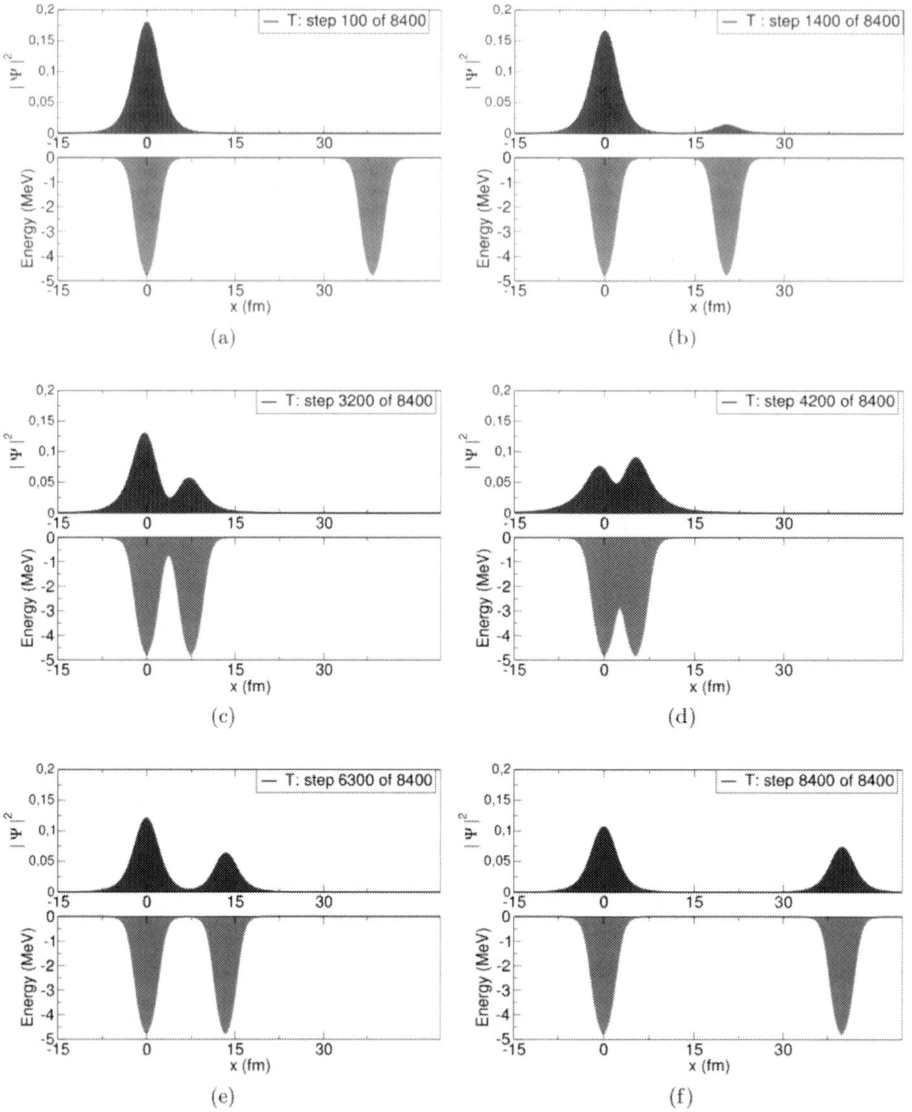


Figure 2: Initially the particle stands in the target potential (a). During the time evolution the projectile approaches (b) and at a certain distance it starts to interact with the target and there is the possibility to transfer the particle (c). When the projectile reaches the distance of closest approach (d), it partially overlaps with the target (the two nuclei do not fuse, this is a peripheral collision). Then the ejectile comes back (e) and when it is far enough one can calculate the transfer probability (f).

Journal of Physics: Conference Series **566** (2014) 012027 doi:10.1088/1742-6596/566/1/012027

## 3. Results

In order to study the different behaviour of bound and weakly–bound nuclei, for the target and projectile potentials I consider here Woods–Saxon wells with same radius, with depths that only admit one bound state. The two depths are chosen in such a way that the first one generates a weakly–bound level, the second a well bound (respectively $V_1$ and $V_2$ in (6)).

$$\begin{array}{ccc} V_1 = -5.00\,MeV & \alpha_T = 0.67\,fm & R = 2.00\,fm \\ V_2 = -12.00\,MeV & \alpha_P = 0.65\,fm & \mu = 0.975\,amu \end{array} \tag{6}$$

The different combinations of $V_1$ and $V_2$ parameters and the different energy levels obtained (at a distance of 40 fm between the two wells) are presented in figure 3.

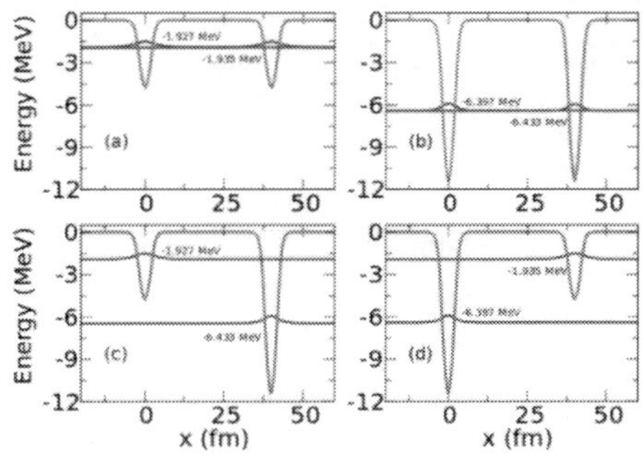


Figure 3: Different combination of potential parameters and corresponding eigenvalues and wave functions. In panel (a) both potentials have depth $V_1$, while in panel (b) both have $V_2$. In panel (c) and (d) they have depths $V_1$ and $V_2$ respectively and vice versa.

### 3.1. Optimum Q–value

The results for the transfer probabilities obtained within the model confirm both theoretical and experimental findings: the optimum Q–value (difference between energy levels) for the transfer of an uncharged particle without change of angular momentum is zero; in fact leaving other parameters unchanged the transfer probability has a maximum when initial and final states in the two wells have the same energy. So for the cases in panels (c) and (d) of figure 3 the transfer probability is highly suppressed, while it is almost maximum in situations like those in panels (a) and (b). In figure 4 the transfer probability is reported as a function of the Q–value.

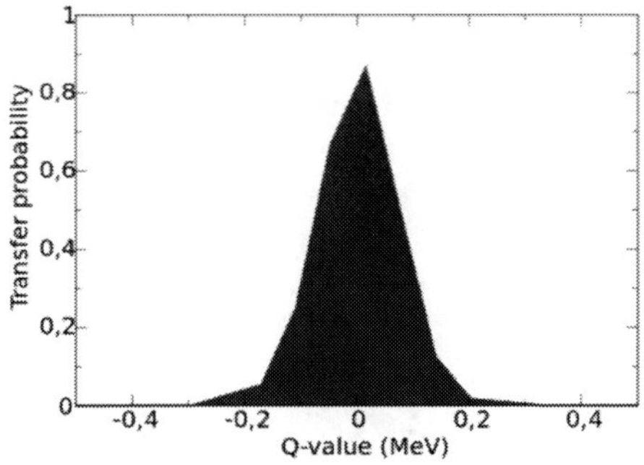


Figure 4: Transfer probability as a function of the Q–value, that is the energy difference between final and initial single–particle states. Optimum Q–value corresponds to zero, where in fact it is found maximum probability.

### *3.2. Bound and weakly–bound systems*

In the optimum Q–value condition, the region corresponding to peripheral collision presents an high transfer probability. For weakly–bound situation transfer occurs more likely, because the wavefunctions extend over a larger range, but excitation to the continuum (break–up reaction) also comes into play. One can in fact observe that as the system becomes less bound, breakup would be more probable than transfer. In figure 5 the transfer probability is reported with respect to the distance of closest approach; the case of two weakly–bound nuclei (like those in figure 3a) is shown on the left, while the bound case (as in figure 3b) is represented on the right side.

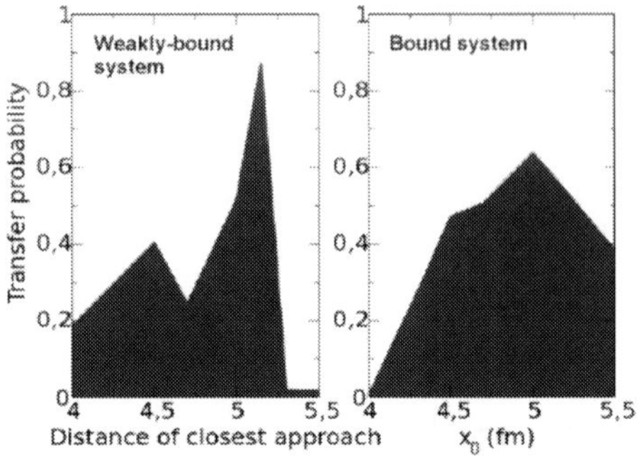


Figure 5: Transfer probability with respect to distance of closest approach, for both bound and weakly–bound systems.

## 4. Conclusions and goals for the future

A one dimensional toy model turns out to be a useful tool even to investigate the properties of rather complex systems. Its simplicity does in fact allow to overcome all those mathematical complexities which usually hinder the full solution. The model encompasses many caracteristics of transfer reactions, such as optimum Q–value and distance of closest approach dependence. The next goal will be to extend the model to two–particle systems in order to describe pair transfer reactions in connection with pairing interaction.

## Acknowledgments

This work has been done in collaboration with Prof. Andrea Vitturi.

## References

[1] Esbensen H, Broglia RA, and Winther A 1982 Comparison of the semiclassical coupled channel theory of particle transfer with exact calculations *Ann. Phys. (NY)* **146** 149–73

[2] Moyer CA 2006 Numerical solution of the stationary state Schroedinger equation using transparent boundary conditions *Comp. Sci. Eng.* **8** 32–40

# AUTHOR INDEX

# AUTHOR INDEX